LIFE SCIENCES RESEARCH AND DEVELOPMENT

A COMPREHENSIVE GUIDE TO APTAMERS

LIFE SCIENCES RESEARCH AND DEVELOPMENT

Additional books and e-books in this series can be found on Nova's website under the Series tab.

LIFE SCIENCES RESEARCH AND DEVELOPMENT

A COMPREHENSIVE GUIDE TO APTAMERS

TOM SHUSTER
EDITOR

nova science publishers
New York

Library of Congress Cataloging-in-Publication Data

ISBN: 978-1-53616-293-6

Published by Nova Science Publishers, Inc. † New York

CONTENTS

PREFACE

This collection opens with a focus on recent advancements on the development of nucleic acid aptamers as alternative delivery systems for therapeutic oligonucleotides. Additionally, key examples of targeted delivery of the most common nucleic acid therapeutics, including small interfering RNAs, short hairpin RNAs, microRNAs and antisense oligonucleotides for a number of disorders are discussed.

The following chapter deals with in vivo studies that were conducted with radiopharmaceuticals based on aptamers, the radionuclides used, the radiolabeling strategies, the chemical modifications of interest to improve their properties, and the main aptamers advantages and drawbacks for application as radiopharmaceuticals.

In conclusion, the authors discuss the perspectives of using aptamers in various detection methods, their advantages and disadvantages, and the results of such work carried out through the present day.

Chapter 1 - Aptamers are highly structured, short single stranded nucleic acids that fold into complex three-dimensional structures that drive their binding to target. As a consequence of these distict structures, their binding is characterized by high affinity and specificity. Aptamers are extremely versatile and can bind to a wide range of targets from proteins, peptides to whole cells and tissues. Recently, aptamers have attracted much attention for their ability to internalize into cells and tissues, and to potentially deliver secondary agents along them, such as therapeutic molecules. This is due to their small size, safer profile (low immunogenicity and toxicity) and exquisite targeting ability on cell surface markers. The following chapter will focus on recent advancements on the development of nucleic acid aptamers as alternative delivery systems for therapeutic oligonucleotides. It will also discuss key examples of targeted delivery of the most common nucleic acid therapeutics, including small interfering RNAs, short hairpin RNAs, microRNAs and antisense oligonucleotides for a number of disorders. Current potential, challenges and ways to address them will also be highlighted.

Chapter 2 - Acid nucleic aptamers are oligonucleotides that bind to a specific target molecule with high affinity and specificity. Because of their unique characteristics, aptamers are promising tools for development of new radiopharmaceuticals. They seem to be non-toxic and non-immunogenic, have small size and fast clearance. Aptamers can be selected for almost any target, including toxic and non-imunogenic molecules. Due to their small size and structural flexibility, aptamers may bind hidden epitopes. They can be easily produced by *in vitro* conditions with high reproducibility and free of contaminants, and the chemical synthesis makes them receptive to many modifications such as to make them more resistant to nucleases or to incorporate chelating groups. Aptamers can be labeled with different radioisotopes, thus allowing its application for imaging, therapy or theranostics, according to the radionuclide used. This chapter deals with the *in vivo* studies that were conducted with radiopharmaceuticals based on aptamers, the radionuclides used, the radiolabeling strategies, the chemical modifications of interest to improve their properties, and the main aptamers advantages and drawbacks for application as radiopharmaceuticals.

Chapter 3 - From the time the first aptamer was discovered with the use of SELEX technology until now the interest of the research community to this topic is growing. The possibility of obtaining oligonucleotide sequences with high affinity to a given target that recognize molecules of various nature and perform it with incredible specificity (including the possibility of obtaining aptamers able to identify small molecules that differ in one functional group), the accessibility of their synthesis and simplicity of making a wide range of modifications give a chance for using aptamers in many applications, even in those still dominated by antibodies. Thus, aptamers are now widely used in various applications: in diagnostic systems (aptaPCR, PCR, biosensors, Lab On Chip, etc.), in therapeutic drugs (Macugen and a lot of drugs at the clinical trials), in research work and routine biotechnology procedures such as affinity chromatographic purification, cytometric analysis, intracellular fluorescence imaging,etc. However, despite the success of many research groups in obtaining RNA and DNA aptamers, they are still not widely used. The practice of selection and modification of aptamers has not yet replaced the adopted practice of obtaining monoclonal antibodies. In this chapter the authors will discuss the perspectives of using aptamers in various detection methods, their advantages and disadvantages, and the results of such work carried out until the present day.

In: A Comprehensive Guide to Aptamers
Editor: Tom Shuster
ISBN: 978-1-53616-293-6

Chapter 1

DEVELOPING APTAMERS FOR THE DELIVERY OF THERAPEUTIC RNA

Styliana Philippou[1], Nikolaos P. Mastroyiannopoulos[1,2] and Leonidas A. Phylactou[1,2,*]

[1]Molecular Genetics, Function and Therapy Department,
The Cyprus Institute of Neurology and Genetics, Nicosia, Cyprus
[2]Cyprus School of Molecular Medicine,
The Cyprus Institute of Neurology and Genetics, Nicosia, Cyprus

ABSTRACT

Aptamers are highly structured, short single stranded nucleic acids that fold into complex three-dimensional structures that drive their binding to target. As a consequence of these distict structures, their binding is characterized by high affinity and specificity. Aptamers are extremely versatile and can bind to a wide range of targets from proteins, peptides to whole cells and tissues. Recently, aptamers have attracted much attention for their ability to internalize into cells and tissues, and to potentially deliver secondary agents along them, such as therapeutic molecules. This

* Corresponding Author's E-mail: laphylac@cing.ac.cy.

is due to their small size, safer profile (low immunogenicity and toxicity) and exquisite targeting ability on cell surface markers.

The following chapter will focus on recent advancements on the development of nucleic acid aptamers as alternative delivery systems for therapeutic oligonucleotides. It will also discuss key examples of targeted delivery of the most common nucleic acid therapeutics, including small interfering RNAs, short hairpin RNAs, microRNAs and antisense oligonucleotides for a number of disorders. Current potential, challenges and ways to address them will also be highlighted.

Keywords: aptamer; therapeutics, DNA, RNA, oligonucleotides, delivery, siRNA, shRNA, microRNA, antisense oligonucleotide

Introduction

Aptamers are highly structured, short, single stranded nucleic acids (usually 20-100 nucleotides long). They tend to fold into unique three-dimensional structures which in turn drives their biding to specific targets with extreme affinity and specificity (Sun et al. 2014; X. Ni, Castanares, et al. 2011). As a consequence of this very distinct molecular recognition of targets, aptamers are extremely versatile and can accommodate binding to a wide range of targets. Furthermore, the target can be rationally chosen against a known protein or you can "blindly" select against more complex targets, such as whole cells, and identify novel or even unknown cell membrane receptors that are only present, or more abundant, in a specific cell type (J. Zhou et al. 2012).

Aptamer-target binding is best viewed as an antibody-antigen interaction. For this reason, aptamers are often referred to as "synthetic antibodies" or "nucleotide antibody analogues" (Shangguan et al. 2008; Daniels et al. 2003). However, as therapeutics, aptamers possess characteristics that outweigh current monoclonal antibodies. The most advantageous is their small size (6-30 kDa, ~2 nm in diameter) that allows them to easily infiltrate cells and tissues (Dassie et al. 2009a). Being at least 10 times smaller than antibodies (150-180 kDa, ~15 nm in diameter), aptamers have a higher tendency to be non-toxic and non-immunogenic, and thus less likely to trigger an immune response in therapeutic applications (White, Sullenger, and Rusconi 2000; Germer, Leonard, and Zhang 2013; A.C. Yan and Levy 2009). In addition, the faster aptamer development time compared to antibodies, the ease of selectivity against a wide range of targets (including compounds that are toxic to animals or non-immunogenic), the production via chemical synthesis that reduces lot-to-lot variability and ensures long term availability with lower synthesis costs, are only some of the additional benefits of aptamers for the manufacturing industry (J. Zhou and Rossi 2017).

The starting point of aptamer selection, is a degenerate library of ~$1x10^{15}$ random single stranded nucleic acids (DNA or RNA, typically chemically modified for enhanced nuclease resistance) (Ellington and Szostak 1990; Tuerk and Gold 1990). Isolation of specific aptamer sequences is then achieved from this library by employing SELEX (systematic evolution of ligands by exponential enrichment). SELEX, in its most basic form, is an iterative process of target binding, partitioning (of binders from non-binders) through washes, recovery and preferential amplification (Hall et al. 2009; A. Yan and Levy 2014; J. Zhou and Rossi 2014a; Mi et al. 2010; Darmostuk et al. 2015). This method is, in essence, an evolutionary process or Darwinian selection where the "fittest" aptamer sequences are captured by the target, thus endure the washing step, while non-binders are discarded (Y. Chen et al. 2004). The bound sequences are separated from the target and preferentially amplified in order to form the next library (or pool) of random sequences. This process is repeated until there is enrichment of specific aptamer sequences (J. Zhou and Rossi 2011).

Aptamer libraries are characterised by high structural complexity and diversity (J. Zhou and Rossi 2011). This feature is key to aptamer selection for any given target and it is owed to the presence of numerous molecules in the starting pool with different base composition. Different sequences then fold to create a library of different structures. Some of these may have the right structure (i.e., shape) to "fit" to a target, and if they survive the selection, they are regarded as potential aptamers for that specific target. Due to this high structural complexity, we are able to select aptamers for almost any target, be that a peptide, a protein, small molecule, a cell or an entire organ/tissue within a living animal (McNamara et al. 2006b). Furthermore, the starting library can accommodate chemical modifications to improve the properties of the selected aptamers, including nuclease resistance and pharmacokinetic profile (S. Ni et al. 2017; Lipi et al. 2016).

SELEX is at the heart of aptamer technology and it can be tailored to meet different applications including biomarker discovery, diagnostic and therapeutic applications. In addition, more steps can be added to the classical selection scheme or even modified, favouring thus the identification of aptamers with particular characteristics that extent beyond the binding of protein receptors. One of the more recent adaptations is the Cell-internalization SELEX that incorporates steps to ensure the selection of aptamers that selectively internalize into target cells upon binding to a membrane receptor (W.H. Thiel, Bair, et al. 2012; K.W. Thiel, Hernandez, et al. 2012b).

CELL-INTERNALIZATION SELEX

In addition to being valuable tools for therapy and diagnosis, the advent of Cell-Internalization SELEX in 2012, has made possible the identification of cell-internalizing aptamers that could potentially act as targeting moieties for the delivery of secondary agents, such as therapeutic molecules (W.H. Thiel, Bair, et al. 2012; K.W. Thiel, Hernandez, et al. 2012b). Similarly to Cell-SELEX, specific cell-surface molecules or even unknown membrane receptors can be directly targeted within their native environment, allowing a straight-forward enrichment of cell-internalizing aptamers (J. Zhou and Rossi 2011). The fundamental difference between Cell-SELEX and Cell-Internalization SELEX is that in the latter both the unbound and surface aptamers are discarded during washes, while only internalized aptamers are recovered (Figure 1). Removal of the surface bound aptamers is a key step in this approach and is achieved by incorporating a stringent wash step with a high salt solution (0.5M NaCl PBS) and/or low pH (0.2N acetic acid) prior to aptamer recovery. Additionally, in order to prevent further receptor-mediated internalization of aptamers, this wash step is performed at a low temperature (4°C), as it is known to inhibit receptor-mediated endocytosis of various molecules, including aptamers (Philippou et al. 2018; Ranches et al. 2017).

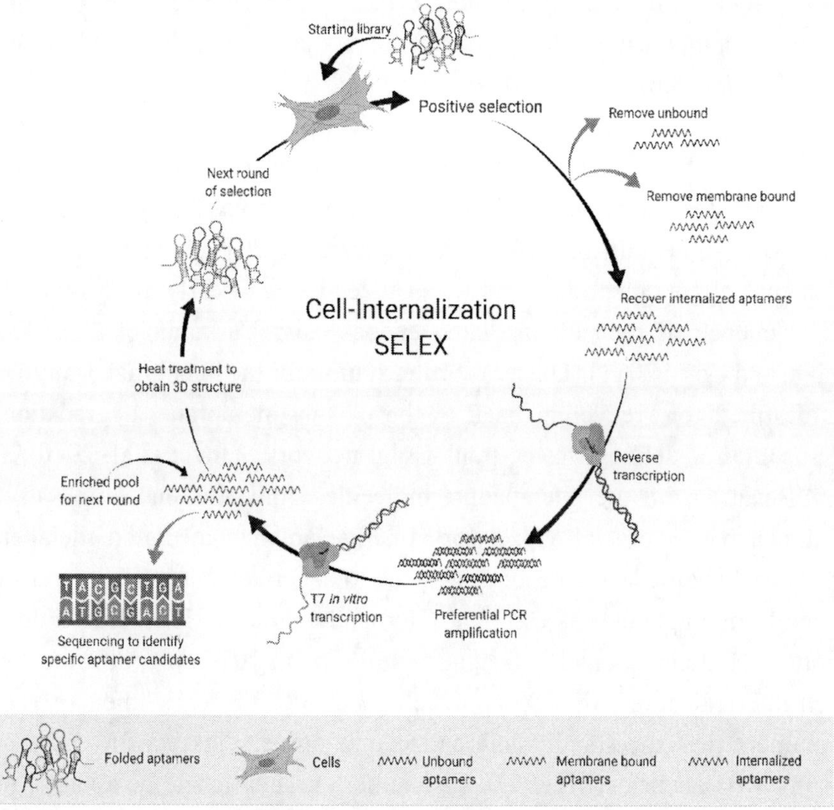

Figure 1. Cell Internalization SELEX process. Aptamer selection strategy. Schematic representation of the evolution of internalizing RNA aptamers for a specific cell target via 15 Cell-internalization SELEX rounds. The random RNA library is incubated with target cells for 15-120 min. The non-binders and membrane bound aptamers are removed through washes, PBS wash and salt wash, respectively. Subsequently, internalized aptamers are recovered via RNA extraction and preferential RT-PCR. Recovered sequences go on to form the next pool of random sequences and this process is repeated until there is enrichment of specific sequences.

The description of this SELEX variation triggered the selection and publication of several internalizing aptamers against many disease-relevant targets. The pioneers of this method first developed an aptamer capable of internalizing into HER2+ mammary carcinoma cells, followed in the same year by another publication of a vascular smooth muscle cell-internalizing aptamer (W.H. Thiel, Bair, et al. 2012; K.W. Thiel, Hernandez, et al. 2012b). This latter study further highlighted the critical importance of utilizing high throughput sequencing with bioinformatics analyses to rapidly identify strong aptamer candidates against complex targets, such as whole cells (W.H. Thiel, Bair, et al. 2012).

While the Cell-Internalization SELEX protocol favours the enrichment of aptamers that selectively internalize through cell surface receptors, it does not guarantee release of the aptamer to the cytoplasm (Hernandez et al. 2013a; J. Zhou and Rossi 2011; Hernandez et al. 2013b). This is due to the fact that oligonucleotides tend to localize first into early endosomes as a consequence of receptor-mediated endocytosis (Varkouhi et al. 2011; J. Zhou and Rossi 2011). This is a sorting station for the internalized cargo that is destined for "recycling back to the plasma membrane, degradation in lysosomes or delivery to the trans-Golgi network" (Jovic et al. 2010). The translocation of the therapeutic molecule from the endosome to the cytoplasm is a crucial aspect for effective aptamer-mediated therapeutic delivery. Therefore, incorporating steps to separate endosome-bound from cytoplasmic aptamers is essential (Hernandez et al. 2013b). Based on this notion, Hüttenhofer and colleagues employed in 2017 an endosome-based cell-internalization SELEX approach to enable the enrichment of DNA aptamers that specifically internalize into target cells via an endosomal pathway (Ranches et al. 2017). The authors first enriched the aptamer pool with cell-bound aptamers and then at the 5th round an additional enrichment step was incorporated, to partition the cell-bound aptamers and select only the endosome-based internalized aptamers. For those aptamers already selected with the classic Cell-internalization SELEX approach, the Giangrande group offers an alternative approach (Law et al. 2017). This group reported a functional assay, termed RNA-RIP, to assess internalization and subsequent cytoplasmic release of the RNA aptamer using a cellular toxin (saporin) with cytotoxic properties. In this particular study, the toxin was conjugated to a prostate-specific membrane antigen (PSMA) RNA aptamer (A9g) and upon localization of saporin to the cytoplasm, the toxin exerted its ribosome inactivating protein effects, leading to cell death. In this way, the cytoplasmic localization of the anti-PSMA A9g aptamer was confirmed (Law et al. 2017; Serfass et al. 2017). Finally, in a more recent study published by our group, while no means were employed to ensure preferential enrichment of cytoplasmic-targeting aptamers, the selected aptamer (A01B) showed minimal endosomal internalization and subsequent cytoplasmic localization (Philippou et al. 2018). This internalizing aptamer is the very first skeletal muscle targeted aptamer to be reported, and its identification could trigger the development of targeted therapies for muscle-related disorders, using A01B as the targeting moiety.

Despite the success of Cell-Internalization SELEX, aptamers selected *in vitro* may not necessarily function *in vivo*. A possible explanation for this difference might be the dependence of aptamers' on their protein target conformation (Mi et al. 2016). The target's conformation is in turn affected by its environment and can vary between *in vitro* and *in vivo* settings. For this reason, a novel method was recently developed to generate tissue-penetrating aptamers directly within living animals(Cheng et al. 2013). This selection method is termed *in vivo* SELEX and rather than using purified proteins or cells, the entire animal is used for developing aptamers (Mi et al. 2016; Mi et al. 2010; Cheng et al. 2013; H. Wang et al. 2018). Researchers administer the RNA library intravenously to the animal, harvest the tissue or organ of interest and extract the bound sequences. This technique was used to identify 2′-fluoropyrimidine-modified RNA aptamers in an animal model of intrahepatic colorectal cancer metastases (Mi et al. 2010). The resulting nuclease-resistant RNAs specifically localize to targets within intrahepatic tumour deposits. In addition, a 2′-fluoropyrimidine-modified RNA library was administered to wild-type mice, and then brains were harvested for the recovery of aptamers that cross the blood-brain barrier via binding to brain capillary endothelial cells (Cheng et al. 2013). These examples demonstrate the efficacy of *in vivo* SELEX as a more direct method for generating RNA aptamers that are suitable for use *in vivo*.

Aptamers as Targeting Moieties

Tissue specificity is a key challenge for systemic delivery of therapeutics (Godfrey et al. 2017). The inability to selectively and specifically target a subset of diseased cells in the body results into off-target side effects, with the most detrimental being the toxicity. Consequently, these drugs are characterized by a poor safety profile and a lower therapeutic index. While there are other routes of administration, for diseases with multiple affected organs/tissues, the systemic route of administration is the only available option. As a result, most of the therapeutic molecule is distributing to liver, kidney and other organs of the excretory system. Eventually, the amount of the molecule that reaches the target is unsatisfactory to induce a therapeutic effect, thus reducing its therapeutic efficacy.

By impacting absorption, distribution, metabolism and elimination (ADME) of a drug, its therapeutic efficacy can be maximized and the side effects minimized (Wen, Jung, and Li 2015). This can be achieved by conjugating the therapeutic molecule with targeting moieties that can assist their entry into cells. Among these, aptamers as the targeting moiety, display a set of characteristics that are superior over other delivery systems, including liposomes, polymers, nanoparticles and cell penetrating peptides (Wen, Jung, and Li 2015; Godfrey et al. 2017; Yoon and Rossi 2018). The high affinity and specificity for their targets is probably the most important advantage of this class of oligonucleotides that renders them useful for targeted delivery. Aptamers form highly specific complexes with their targets that enable them to distinguish even between closely related molecules, such as conformational isomers or a protein target with a single amino acid changes (Geiger et al. 1996; L. Chen et al. 2015). As a result of these tight associations with the target, aptamers are characterised by very low dissociation constants, between low picomolar and nanomolar, a feature that makes them also ideal molecules for therapeutic and diagnostic applications (O'Sullivan 2002; Zhang, Blank, and Schluesener 2004). Furthermore, their small size is beneficial for smaller targets or binding domains which are hidden and inaccessible to antibodies. In addition, as a consequence of their synthetic nature, aptamers are easily chemically modifiable during SELEX. This avoids the risk of altering aptamer folding that is necessary for specific binding when chemical modifications are added post selection (Lin et al. 1994; Burmeister et al. 2005). The most commonly used chemical modifications are sugar modifications with fluoro- (F), amino- (NH2) or O-methyl (OCH3) groups, or by using locked nucleic acids (LNA) and these are typically introduced at the 2'-position of the nucleotide. Backbone modifications can also be introduced, such as hydrophobic groups, phosphorothioates, amino acids and many more (S. Ni et al. 2017; Lipi et al. 2016). All these modifications result into aptamers with greater functional diversity and improved resistance against serum nucleases, both indispensable features for a targeted drug delivery system.

The following chapter will review key studies that have set the grounds for the use of aptamers as cell-specific targeting moieties for therapeutic delivery. We will highlight recent advancements in aptamer-guided delivery of RNAi conjugates in particular. Finally, we will address current challenges in therapeutic delivery guided by aptamers along with potential solutions.

Aptamers as Targeting Moieties for RNAi

RNA Interference

Many organisms use RNAi to control genes and this biological process can also be used as a tool in the laboratory and in the future perhaps as a therapy. Eukaryotic cells have many sophisticated ways of controlling gene expression. In the complex environment of a cell, these processes need to be precisely targeted. For example, there is a group of mechanisms that use small regulatory RNA molecules to direct gene silencing and this is called RNA interference (RNAi). There are several types of small regulatory RNAs. Small interfering RNA (siRNA), microRNA (miRNA), short hairpin RNA (shRNA) and more recently antisense oligonucleotides (AONs) are some of the key players in RNAi. These regulatory RNA can have their silencing effect, primarily into the cytoplasmic region of a cell, by targeting an mRNA for degradation, suppression of translation or by inducing alternative splicing of genes (Juliano et al. 2008).

RNAi therapies are an exciting technological platform providing a feasible alternative when all other therapeutic approaches fail. However, RNAi therapeutics still face challenges that outweigh their promise as personalized, targeted therapies (Juliano et al. 2008). Perhaps, the most critical issue concerns the effective delivery of the RNAi therapeutics to specific diseased cells and organs.

The delivery challenges have inspired innovation in biotechnology with regards to drug delivery. As a result, more emphasis is given on the development of RNAi delivery systems to provide active transport of the therapeutic on site. Aptamer-guided RNAi therapeutics have been in focus in the research community for quite a while, with several outstanding examples in the preclinical settings. To date, a wide range of molecules have been conjugated to aptamers for targeted delivery, including chemotherapeutic agents (primarily doxorubicin), siRNA, miRNA, peptides, or even other aptamer (Pastor et al. 2011; Schrand et al. 2014; Schrand et al. 2015; C. Li et al. 2016; Yoon et al. 2016). Such delivery systems are able to specifically bind to target cells and selectively deliver their cargo within the cell. The following sections will discuss selected examples of aptamer-guided delivery of the main RNAi players: siRNA, miRNA, shRNA and AON, in the most active research field to date- cancer therapy and infectious diseases.

Aptamer- siRNA Delivery

Small interfering RNAs is one of the key RNAi effector molecules. Small interfering RNAs known as siRNAs, derived from longer double stranded RNAs that are either produced in the cell itself or are synthetically produced and delivered into cells experimentally. The introduction of siRNA or double stranded RNA is widely used to manipulate gene expression (Figure 2A) (Juliano et al. 2008). However, as therapeutics, siRNAs face several challenges, including poor stability due to nucleases present in serum and lack of a specific-cell targeted delivery mechanism to direct the siRNA on site. Towards this end, aptamers hold the promise of being effective delivery tools for several therapeutic RNA molecules including siRNAs.

Cancer, is a primary target for siRNA therapeutics and as a consequence, for aptamer-siRNA conjugates as well. The examples are numerous, with aptamer-mediated siRNA delivery showing potential for many types of cancers. Prostate cancer and more specifically the prostate specific membrane antigen (PSMA) receptor, was the first cancer-related receptor explored for siRNA delivery due to its ability to selectively internalize in prostate cancer cells (Farokhzad et al. 2004). This internalizing aptamer was selected with Cell-SELEX, long before the discovery of Cell-Internalization SELEX, hence the immense interest in its use as a taregeting moiety. In a proof-of-concept study, Chu et al. generated a highly specific aptamer-streptavidin-conjugated siRNA chimera that was rapidly taken up by PSMA positive cells (LNCaP) followed by fast inhibition of gene expression (Chu et al. 2006). This chimera, composed of the A10 anti-PSMA RNA aptamer and the anti-laminin A/C siRNA or GAPDH siRNA, was able to attain inhibition similalry to lipid based transfection (Figure 3D). The high inhibition was further assisted by the incorporation of disulphide linkage between the streptavidin core and the biotinylated siRNA that cleaved upon entering the reducing environment of the cell, releasing the siRNA form the bulky substrate (streptavidin; interferes with siRNA processing) (Chu et al. 2006). In addition, as the streptavidin is a tetramer with four functional binding sites for biotin, the therapeutic siRNAs carried by an aptamer-streptavidin targeting moiety could have multiple targets, demonstrating a new venue for therapeutic delivery of siRNAs characterized by cell type-specific gene silencing effect.

Despite the ease of preparing an aptamer-streptavidin-siRNA conjugate, McNamara et al. explored the possibility of generating an all RNA chimeric molecule for easier synthesis as well as RNAi processing (McNamara et al. 2006a). Using the same A10 anti-PSMA aptamer the authors covalently conjugated the aptamer with the polo-like kinase 1 (Plk1) and B-cell lymphoma 2 (Bcl-2) siRNAs (Figure 3A, siRNA). Both PLK1 and Bcl-2 are survival genes overexpressed in most tumours making them good targets for siRNA-mediated silencing (Eckerdt, Yuan, and Strebhardt 2005; Cory and Adams 2005). Combined, the A10-Plk1 and the A10-Bcl2 chimeras were able to induce apoptosis to LNCaP cells but not in PC-3 cells (a distinct prostate cancer cell line not expressing the cell-surface receptor PSMA), demonstrating also the ability of the chimera to escape the endosome with subsequent localization in the cytoplasm. Furthermore, when A10-Plk1 was intratumorally injected in a PSMA positive xenograft model of prostate cancer, inhibition of tumour growth was also observed. Notably, the aptamer-mediated nature of the gene silencing was demonstrated by generating a chimera with mutations in the PSMA binding site. This mutant chimera lost its binding activity completely (McNamara et al. 2006a).

In the following years, the A10 anti-PSMA aptamer served as a key example to elegantly illustrate the versatility, robustness of aptamers, and the ability to further optimize the design and binding affinity of aptamers even after selection. For example, in a follow-up study by the same group, the authors attempted to improve the efficacy and siRNA silencing efficiency for systemic administration (Dassie et al. 2009b). This feature (i.e., intravenous administration) is crucial for the clinical translation of the chimera. One of the most notable modifications to the A10 aptamer was the sequence modifications made to favour loading of the guide siRNA strand onto the Dicer and to increase the stability of the duplex (Figure 3G). Additionally, the circulation half-life of the chimeras' was enhanced by conjugation to a polyethylene glycol (PEG) molecule whereas the size of the aptamer (A10) was reduced from 71 nucleotides to 39 nucleotides (A10-3.2), making it more amenable to large-scale chemical synthesis. The binding affinity of the shortest truncated version was comparable to the full-length aptamer whereas its antitumor activity for prostate cancer cells expressing the PSMA receptor was higher when administered systemically. In addition, a lower therapeutic dose of A10-3.2-Plk1 was able to drastically reduce the percentage of mice with metastasis as well as the number of metastases per mouse (Dassie et al. 2009b). This indicated that the aptamer acts as an inhibitor, blocking *de novo* metastases from forming.

Additional shorter anti-PSMA aptamers were also generated following the rational truncation approach described by Rockey et al. (Rockey et al. 2011). The aim of this study was to provide a rational approach to replace the traditional truncation via a trial-and-error process with a more precise and efficient method guided by computational structural modelling, such as RNA structure prediction and protein/RNA docking algorithms. Rational truncation of the A9 aptamer, another PSMA targeting aptamer, was able to produce a truncated version (A9g, 43 nt) that retained a binding profile similar to the full-length A9 (70 nt). As noted in this study, rational truncation should be complemented with selective base changes (5', 3' or internally) in order to maintain the T7 transcription start site (5'GGG) as well as any necessary base-paring complementarity at the ends (Rockey et al. 2011). The approach described herein can be universally applied to identify the minimum functional versions of other aptamers for conjugation of therapeutics and more importantly clinical translation.

Continuing down the line of rational designs, the study by Wullner et al. demonstrated enhanced anti-PSMA aptamer-guided Eukaryotic Elongation Factor 2 (EEF2) induced-cytotoxicity, and ultimate apoptosis of prostate cancer cells by creating a bivalent aptamer-siRNA chimera (Wullner et al. 2008). EEF2 is a key component of the translational machinery that if supressed, leads to inhibition of protein synthesis and in turn induction of apoptosis and cell death (Jorgensen, Merrill, and Andersen 2006). In this study, the efficacy of the aptamer-siRNA chimera was enhanced by using two A10 anti-PSMA truncated aptamers, xPMS-A10-3, rather than one (Figure 3F). The EEF2 siRNA portion was inserted into the rigid spacer separating the two aptamer, allowing in this manner proper folding of the aptamers into their active conformation (Wullner et al. 2008). Although the bivalent aptamers had greater affinity and specificity for LNCaP cells, as well as internalization ability (4 times more than their monovalent counterparts) they were unable to show improvement in the silencing effect when compared to the monovalent aptamer-siRNA construct. In a more recent study, Liu et al. created another bivalent anti-PSMA aptamer but this time for the delivery and silencing of two different genes, the epidermal growth factor receptor (EGFR) and the gene survivin (Figure 3E) (H.Y. Liu et al. 2016). EGFR overexpression is associated with increased cancer proliferation, tumour vascularization and metastasis whereas survivin is a key member of the inhibitor of apoptosis protein family, with critical role in the progression of prostate cancer and other tumours (Howe and Brown 2011; Altieri 2013). Each aptamer was covalently conjugated to one siRNA and then the two portions were linked together with a stretch of uracil bases ("UUUU"). The bivalent aptamer-dual siRNA chimera effectively knockdown both EGFR and survivin in prostate cancer cells *in vitro* and effectively supressed tumour growth and angiogenesis in a xenograft mouse model of prostate cancer. A different conjugation approach was proposed by Ni et al. in 2015 (X. Ni et al. 2015). In an attempt to overcome the limitations of *in vitro* transcription, the anti-PSMA aptamer A10-3.2 was conjugated to the siRNA via utilizing a region on the 3' end of the aptamer (not involved in aptamer binding) as a bridging region for chimera assembly (Figure 3C). Complementary sequences were next incorporated to the sense strand of the siRNA, allowing the conjugation of the therapeutic molecule with the aptamer via a simple annealing process. This mimics the universal "sticky bridge" conjugation approach published in earlier years by Zhou et al. (J. Zhou et al. 2009a). With this approach, the authors were able to selectively deliver the chimera to prostate cancer cells and to induce targeted radiosensitization in a mouse xenograft model, enhancing their potency to radiation therapy (X. Ni et al. 2015).

Another key challenge for the delivery of siRNAs is their stability *in vivo* (M. Kim et al. 2018). In addition to nucleic acid chemical modification strategies, protection against serum nucleases can be achieved by encapsulating siRNAs in nanocarriers, with further attachment to an aptamer for targeted delivery (reviewed in detail in (K. Chen et al. 2017; M. Kim et al. 2018)). For example, Wu et al. reported in 2018, a novel anti-PSMA nanostructure loaded with a siRNA that silences the expression of the Forkhead box M1 transcription factor (FoxM1); an important antitumor target involved proliferation and overexpressed in prostate cancer tumours (M. Wu et al. 2018; Y. Wang, Yao, et al. 2014). The construct was composed of lipid-based nanobubbles as a carrier for the siRNA, that were subsequently conjugated to the truncated anti-PSMA aptamer, A10-3.2, for cell targeting. Upon ultrasound, these ultrasound-responsive nanobubbles were able to access the cell membrane followed by siRNA release in prostate cancer cells followed by a high gene silencing effect. *In vivo*, the nanostructure led to significant inhibition of tumour growth and prolonged the survival of mice, due to reduction in FoxM1 expression, and a higher apoptotic activity (M. Wu et al. 2018). Contrasting to this work, Bagalkot et al. made use of a simpler nanocarrier, polyethylenimine (PEI) -coated quantum dots, to induce GFP silencing in PSMA-positive C4-2B cell *in vitro* (Figure 3I) (Bagalkot and Gao 2011). In addition to their role as the core of the nanocarrier, the quantum dots, which are widely applied as fluorescent probes, served as intracellular monitoring tools via fluorescent imaging (Figure 4).

In the study by Xu et al. a different technology was developed for linking aptamer-siRNA chimeras to nanocarriers (Xu et al. 2017). Specifically, a PDPA formulation, which stands for oligoarginine-graft PEF-[poly(2-(diisopropylamino)ethyl methacrylate)], was used to deliver and silence the effect of Prohibitin 1, a gene that is overexpressed in PSMA-positive prostate cancer, *in vitro* and *in vivo,* utilizing a xenograft mouse model. Importantly, the formulation of the aptamer-siRNA chimera with fusogenic oligoarginine peptides permitted enhanced endosomal escape with a subsequent increase in siRNA bioavailability to target. A similar effect can be obtained with polyethyleneimine (PEI) polymer-coated quantum dots with an equally high endosomal escape (Bagalkot and Gao 2011). In addition, this study stressed the fact that controlled orientation of the aptamer-siRNA can dramatically enhance the gene-silencing effect of the nanostructure. Furthermore, similar to Bagalkot et al. the use of quantum dots had a dual role: serving as the core of nanostructure at first and secondly as a fluorescent imaging tool (Bagalkot and Gao 2011). The latter is one of the several advantages of using aptamer-conjugated nanocarries for siRNA delivery rather than the traditional one aptamer-one siRNA assembly (Sivakumar et al. 2019). Lastly incorporating aptamer-siRNA chimeras in nanostructures can potentially improve the pharmacokinetic properties of the chimera, an area that is still problematic due to the small size of the aptamer and its conjugates.

Another thoroughly investigated aptamer, is the AS1411 nucleolin aptamer. Nucleolin is a protein abundant in the nucleus that is also overexpressed on the cell membrane of several malignant cells (Z. Chen and Xu 2016). This unique property, in combination with a high internalization capacity has triggered the use of the nucleolin aptamer as a carrier molecule for siRNA-mediated gene silencing for various cancers, including prostate cancer, breast cancer, gastric cancers, lymphocytic leukemia and melanomas (Otake et al. 2007; Qiu et al. 2013; L. Li et al. 2014). For lung cancer in particular, in an attempt to control early metastasis which is a major cause of morbidity in patient with lung cancer, the nucleolin aptamer (aptNCL) was conjugated with the SLUG and NRP1 siRNAs (Lai et al. 2014). The snail family zinc finger 2 (SLUG) and neuropilin 1 (NRP1), when overexpressed, they promote malignant transformation and activate signalling pathways associated with lung cancer metastasis (Shih et al. 2005; Hong et al. 2007). Using a combined treatment of aptNCL-SLUG siRNA chimera and aptNCL-NPR1 siRNA chimera, these key signalling pathways were selectively blocked in lung cancer cells. As a consequence, a synergistic inhibition of lung cancer cell invasion, tumour growth and angiogenesis was observed in a xenograft mouse model. On the contrary, in order to induce a selective cytotoxic effect on melanoma cells A375 as a potential treatment for malignant melanomas, Li et al. encapsulated the anti-BRAF siRNA in PEGylated liposomes guided by the AS1411 nucleolin aptamer for targeted siRNA delivery (L. Li et al. 2014). A bifunctional PEG linker was used in this study to covalently link the cationic liposome with the aptamer, providing the additional benefit of improved pharmacokinetic profile due to the size increase above the cut-off limit of kidneys (Figure 3H). Furthermore the results of this study illustrated the potential of this nucleolin-based nanoconstruct as a therapy for melanomas. Another versatile platform for breast cancer was suggested recently by Wang et al. utilizing again the nucleolin aptamer (Y. Wang et al. 2017). The platform was composed of (i) the nucleolin aptamer AS1411 for cell targeting- as it is highly expressed also in breast cancer cells, (ii) the extracellular vesicles as the nanocarrier and (iii) the vascular endothelial growth factor (VEGF) siRNA or the let-7 miRNA as the therapeutic molecules. Furthermore, cholesterol conjugation was employed to link the nucleolin aptamer on the surface of the extracellular vesicles which are rich in lipids, predominantly cholesterol. The results of this study revealed inhibition of triple-negative breast cancer cells (oestrogen negative, progesterone negative and HER2+ negative) growth both *in vitro* and *in vivo* with no noticeable toxicity or immunogenicity (Y. Wang et al. 2017). Furthermore, the aptamer-modified extracellular vesicle approach could be used useful in other targeted strategies as well.

A very recent publication by Yang et al. demonstrates once again the robustness of aptamers when conjugated with nanomaterials such as functionalized polymeric micelles (Yang et al. 2018). The aim of this study was to evaluate the ability of these micelles for codelivery of Doxorubicin (Dox, a chemotherapeutic drug), along the Toll-like receptor 4 siRNA as a potential lung cancer treatment. The synthesis of these micelles, required a number of different elements, including a hydrophobic core made of urocanic acid for Dox loading, a chitosan-based PEI conjugated polymer to act as the carrier of the therapeutics, disulphide bonds to weaken the powerful positive charge of PEI polymer and the aptamer AS1411 as the targeting moiety to induce specific delivery and reduce potential toxic effects from the micelles. The *in vitro* results revealed sustained release of the siRNA and Dox and improved cytotoxicity in A549 lung cancer cells. Similar results were obtain following systemic administration in mice with high tumour cytotoxicity, suppression of invasion and low toxicity, suggesting a novel therapeutic approach for lung cancer treatment (Yang et al. 2018). The possible use of micelle-based nanoconstructs for other diseases is further suggested through this study.

Using the anti-PSMA aptamer and the nucleolin aptamer as examples, we have described a number of different approaches for aptamer conjugation to siRNAs for the purpose of targeted delivery.

Table 1. Summary of aptamer-siRNA chimeras in Cancer Therapy

Aptamer target	Aptamer formulation	Target gene (or protein)	Type of malignancy	Outcome	Reference
HER+2	Aptamer-siRNA chimera	Bcl-2	Breast cancer	HER2 aptamer-Bcl-2 siRNA conjugates selectively internalize into HER2 positive cells and silence Bcl-2 gene expression. Bcl-2 silencing sensitizes these cells to chemotherapy (cisplatin).	Thiel et al. 2012 *(Pioneers of Cell-internalization SELEX)*
	Three-in-one aptamer-siRNA chimera	EGFR siRNA, HER2, and HER3	Breast cancer	Enabled down-modulation of the expression of all three receptors (HER2, HER3, EGFR), triggering apoptosis of HER2+ breast cancer cells, both *in vitro* and *in vivo*.	Yu et al. 2018
	Bivalent HER2 aptamer-EGFR siRNA aptamer chimera	EGFR	Breast cancer	Aptamer-siRNA chimera selectively taken up by cancer cells expressing HER2+ *in vitro* and *in vivo*. Triggered cell apoptosis, decreased HER2 and EGFR expression, and suppressed tumor growth in breast cancer mouse xenografts.	Xue et al. 2018
	DOTAP-PLGA-PEG	P-gp	Breast cancer	Efficient inhibition of P-gb in Her2+ breast cancer cells *in vitro*.	Powel et al. 2017

Aptamer target	Aptamer formulation	Target gene (or protein)	Type of malignancy	Outcome	Reference
4-1BB	Aptamer-siRNA chimera	mTOR complex 1	Melanoma	Inhibition of mTOR complex 1 signalling in circulating CD8+ T cells and generation of potent CD8+ T cell memory.	Berezhnoy et al. 2014
	Aptamer-siRNA chimera	IL-2R alpha (also known as CD25)	-	Systemic administration of the 4-1BB aptamer-CD25 siRNA conjugate downregulated CD25 mRNA only in 4-1BB-expressing $CD8^+$ T cells promoting their differentiation into memory cells. Treatment with the 4-1BB aptamer-CD25 siRNA conjugates enhanced the antitumor response of a cellular vaccine or local radiation therapy	Rajagopalam et al. 2017
	Aptamer-siRNA chimera	Smad 4	Breast cancer	4–1BB aptamer -Smad4 siRNA potentiates vaccine-induced antitumor immune response *in vivo*.	Puplampu et al. 2018
CD133	Aptamer-PLGA nanoparticles	Salinomucin	Hepatocellular carcinoma cells	Promotion of salinomucin delivery to CD133+ hepatocellular carcinoma cells, with subsequent induction of apoptosis.	Jiang et al. 2015

Table 1. (Continued)

Aptamer target	Aptamer formulation	Target gene (or protein)	Type of malignancy	Outcome	Reference
CD133	Aptamer-siRNA chimera with PEG	Adenosine kinase (ADK)	Epithelial progenitor cells	Selective delivery of chimera in CD133+ cells *in vivo*.	Cheng et al. 2015
EpCAM	Aptamer-siRNA chimera	Survivin	Breast cancer	Survivin silencing sensitized doxorubicin-resistant cancer stem cells to doxorubicin in a mouse xenograft model. Additionally, there was reversal of chemoresistance, suppression of tumour growth and prolonged survival in mice with chemoresistant tumours.	Wang et al. 2015
	Aptamer-siRNA chimera	PLK1	Breast cancer	Targeted gene silencing in breast cancer cells *in vitro*.	Gilboa-Geffen et al. 2015
	PEI	EpCAM	Breast cancer	Inhibited the cell proliferation of MCF-7 and WERI-Rb1 cells *in vitro*.	Subramanian et al. 2015
	Bispecific aptamer fused together with a double stranded RNA adaptor	CD44 and EpCAM	Ovarian cancer	Supressed both CD44 and EpCAM genes with subsequent inhibition of cancer cell growth *in vitro*. Additionally, it effectively inhibited intraperitoneal ovarian cancer growth in a mouse xenograft model.	Zheng et al. 2017

Aptamer target	Aptamer formulation	Target gene (or protein)	Type of malignancy	Outcome	Reference
EpCAM	PEI-SWNT (single-walled carbon nanotube)	BCL91	Breast cells	Efficient cellular uptake of siRNA and suppression of BCL91 in breast cancer cells.	Mok and Park (2012), Mohammadi et al. 2015
MUC1	Dox-aptamer-siRNA chimera	Bcl-2	Breast cancer	Efficient cytotoxic effect on MDR breast cancer cells	Jeong et al. 2017a
	PEI/Holiday junction loaded with multiple aptamer-siRNA chimeras	GFP	Breast cancer	High gene silencing efficiency by the Holiday junction loaded with multiple chimeras compared to a single aptamer-siRNA chimera. Efficient gene silencing in mucin-1 positive KB and MCF-7 cells *in vitro.*	Jeong et al. 2017b
	Multivalent comb-type aptamer-siRNA	Bcl-2	Breast cancer	Reduced cell proliferation in mucin-1 positive cells MRF-7, *in vitro*.	Yoo et al. 2014
CD44	PEG liposome	Luciferase 2	Breast cancer	Gene silencing in CD44+ breast cancer cells *in vitro* and *in vivo*.	Alshaer et al. 2018
	Bispecific aptamer fused together with a double stranded RNA adaptor	CD44 and EpCAM	Ovarian cancer	Supressed both CD44 and EpCAM genes with subsequent inhibition of cancer cell growth *in vitro*. In addition, it effectively inhibited intraperitoneal ovarian cancer growth in a mouse xenograft model.	Zheng et al. 2017

Table 1. (Continued)

Aptamer target	Aptamer formulation	Target gene (or protein)	Type of malignancy	Outcome	Reference
"Orphans"					
EGFR (EGF receptor)	Aptamer conjugated liposomes containing quantum dots and siRNA molecules	Bcl-2	Breast cancer	Targeted gene silencing in EGFR-positive epithelial, human breast cancer cell line. Also, selective accumulation in breast cancer mouse xenografts. Additionally, conjugation to quantum dots offered the ability to fluorescently track the construct *in vivo*.	Kim et al. 2017
PDGFRβ (platelet – derived growth factor β)	Aptamer-sticky bridge-siRNA	STAT3	Glioblastoma	Reduction in cell viability and migration *in vitro* and inhibition of tumour growth and angiogenesis *in vivo* in a mouse xenograft model.	Esposito et al. 2018
αvβ3 integrin	Aptamer-siRNA chimera	EEF2 (eukaryotic elongation factor 2)	Glioblastoma, cervical and prostate cancer	Effective silencing of the EEF2 gene, inhibition of cell proliferation and induction of apoptosis in cancers expressing αvβ3 integrin.	Hussain et al. 2013
TfR(transferrin receptor)	Folate-PEG-conjugated chitosan nanoparticles	*c-Myc*	Breast cancer	High accumulation of siRNA and efficient tumor inhibition in breast cancer xenografts in mice	Li et al. 2017

Aptamer target	Aptamer formulation	Target gene (or protein)	Type of malignancy	Outcome	Reference
Other key publications					
Anti-hTfR RNA aptamer and C10.36 DNA aptamer	Three-way junction (3WJ) nanostructure	-	B cell lymphoma (aptamer C10.36 binds to human Bcell cancer cell line via an unidentified surface antigen that also overexpresses hTfR	A modular nanostructure for cellular delivery of large, functional RNA payloads (50–80 kDa, 175–250 nt) by aptamers that recognize multiple human B cell cancer lines and transferrin receptor-expressing cells. Fluorogenic RNA reporter payloads enable accelerated testing of platform designs and rapid evaluation of assembly and internalization. Modularity is demonstrated by swapping in different targeting and payload aptamers.	Porciani et al. 2018
41t, against platelet-derived growth factor (PDGF), TE17 and sgc8c	Three-dimensional DNA origami box	-	-	An autonomous DNA nanorobot capable of transporting molecular payloads to cells, sensing cell surface inputs for conditional, triggered activation, and reconfiguring its structure for payload delivery.	Douglas et al. 2010

In a similar manner, aptamers for several other types of cancers have been developed, utilizing other well-known cell surface receptors as targets. Examples of these include the epidermal growth factor (EGFR) expressed in breast cancer, lung cancer and gliomablastoma, human epidermal growth factor receptor 2 (HER2+) overexpressed in breast cancer and platelet derived growth factor β (PDGFβ). Additionally, different siRNAs have been delivered to target oncogenes and/or genes involved in downstream cancer-related cellular pathways or biological processes such as STAT3, mTORC, IL-2R alpha and smad4. Similarly to the anti-PSMA and nucleolin aptamer, aptamer-functionalized nanostructure have been developed through conjugation with various nanomaterials. A non-exhaustive list covering other cancer-related aptamer advancements for siRNA delivery is presented in Table 1.

Apart from cancer, aptamer-guided siRNA delivery has been utilized as an alternative therapy for Human Immunodeficiency virus-1 (HIV1) as well. Rossi and colleagues are responsible for most innovations in this field with their research focusing primarily on glycoprotein gp120, as the aptamer target, for the synthesis of anti-viral-aptamer-siRNA chimeras (Jiehua Zhou and Rossi 2014b). The gp120 glycoprotein is an envelope glycoprotein on HIV-1 infected cells playing a key role in viral entry into host cells. In two earlier studies, Rossi and colleagues investigated the potential of the anti-gp120 RNA aptamer to selectively bind to gp120 glycoprotein expressing HIV-1 infected cells with subsequent internalization. In the first study covalent conjugation (via a uracil base, "UU," linker) was employed to join the 3' end of the aptamer to the sense strand of an anti-tat/rev siRNA whereas in the second study the authors introduced a complementary "sticky bridge" GC-rich sequence (17 nts) between the aptamer and the therapeutic siRNA (J. Zhou et al. 2008b; J. Zhou et al. 2009a). This universal "sticky" sequence was also comprised of a three-carbon linker allowing attachment of various siRNA substrates onto the aptamer (Figure 3B). Since then, the "sticky bridge" linker has been used by many other researchers for aptamer-mediated delivery of therapeutics (J. Zhou, Rossi, and Shum 2015; J. Zhou et al. 2013). In both studies, the anti-gp120 aptamer provided selective binding and internalization into infected cells followed by inhibition of HIV-1 replication due to siRNA delivery to the cytoplasm, as well as from the anti-HIV activity of the aptamer itself (acts as an antagonist for the gp120 receptor). Furthermore, in a third study by the same group, the ability of this aptamer-siRNA chimera to supress HIV-1 viral loads and to protect humanized mice from helper CD4(+) T cells was further verified (Neff et al. 2011b). In the 2013 publication, Rossi and colleagues utilized the same aptamer-"sticky bridge" strategy but this time for the delivery of three different siRNAs *in vivo,* in the form of a cocktail treatment (J. Zhou et al. 2013). The reasoning behind this approach was to mitigate viral escape mutants that are often observed in HIV-1 resulting into RNAi-based HIV-1 therapeutic approaches often being ineffective over time. Intravenous weekly injections with this aptamer-siRNA chimera cocktail was able to downregulate all three targeted transcripts with subsequent suppression of HIV-1 viral loads and protection against CD4+ cells, being thus in agreement with their previous results as well (Neff et al. 2011b). In a similar manner, an aptamer that selectively recognizes CD4+ T cells, was conjugated with a siRNA against HIV *gag/vif* or C-C chemokine receptor type 5 (CCR5), demonstrating once again effective inhibition of HIV infection *in vitro* and *in vivo* (Wheeler et al. 2011). A CCR5 targeted RNA aptamer was also developed by Zhou e*t al.* in 2015 (J. Zhou et al. 2015). CCR5 is regarded as a key player for viral entry and an attractive cellular target for the treatment of HIV-1 (Jiehua Zhou and Rossi 2014b). This aptamer was shown to successfully target HIV-1 susceptible cells, to specifically regulate HIV-1 gene silencing and block the CCR5 receptor required for HIV-1 entry in host cells. The same group recently reported the first example of an aptamer that can deliver a promoter-targeted small RNA capable of inducing transcriptional gene silencing (TGS). Using this approach, infected cells were targeted for stable HIV-1 silencing (J. Zhou et al. 2018). This aptamer can also act as a carrier molecule for additional therapeutic RNAs to epigenetically silence other genes involved in disease.

Two more studies are worth mentioning in this section. Firstly, the work published by Sánchez-Luque *et* al. focusing on the development of the smallest in size HIV-1 RNA aptamer, a 16 nucleotide long, that targets the 5'untralated region of the HIV-1 genome (Sanchez-Luque et al. 2014).

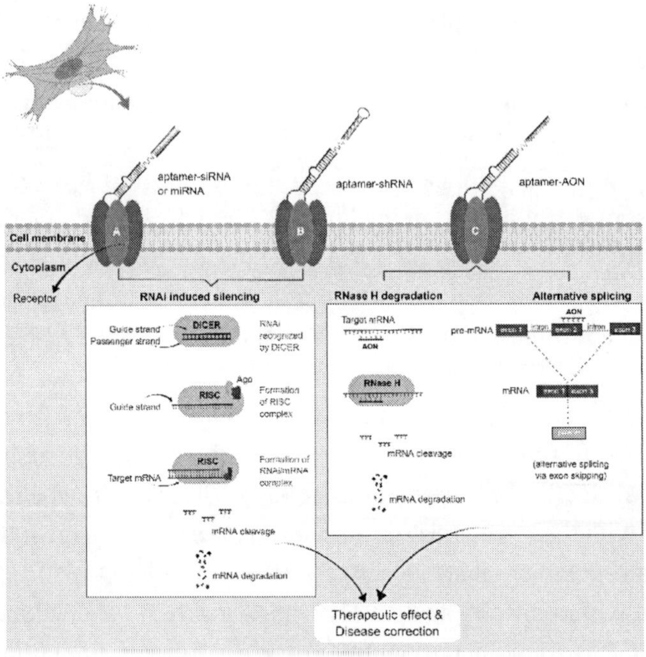

Figure 2. Receptor-mediated internalization of aptamer-RNAi conjugates. Depending on the therapeutic sequence being internalized, different intracellular RNAi mechanisms are activated: A. aptamer-siRNA or miRNA, B. aptamer-shRNA and C. aptamer-AON. For A and B, the precursors of siRNA, miRNA and tha shRNA bind to DICER, which is an endonuclease protein that cuts the RNA into short segments. Most siRNAs and microRNAs are approximately 21 nucleotides long. The short double stranded RNA then binds an Argonaute protein (Ago). One strand of the RNA is selected and remains bound to the Argonaute, this is called the guide or sense strand whereas the passenger strand or antisense is degraded. The combination of the RNA and the Argonaute along with other proteins is called the RNA induced silencing complex or RISC. siRNA, miRNA and shRNA direct RISC to bind to specific messenger RNA (mRNA) in a precise manner, because it is determined by base pairing between the therapeutic RNAi sequence and the target mRNA. Once bound to their mRNA targets, Argonaute catalyses cleavage of the mRNA which will then be degraded. AONs on the other hand exert their function through different mechanisms, including RNase H degradation and alternative splicing. RNase H is an endogenous enzyme that cleaves the RNA strand of the RNA-DNA duplex. This is the most widely mechanism for AON-mediated gene knockdown. The AON binds to the target mRNA via base pair complementarity. The RNase H enzyme, recognizes the duplex and induces cleavage and degradation of the mRNA (left). In the exon skipping approach, the mutant exon is musked by an AON that is directed against it. During splicing, this exon is skipped leading to restoration of the open reading frame (ORF) and production of a truncated but functional protein (right). The latter is a form of alternative splicing that induced disease correction via alternative mRNA splicing.

Herein, the authors used an *in vitro - in silico* combined approach to rationally guide the design and optimization of this aptamer, termed RNApt16, which is a highly active inhibitor of HIV-1. Secondly, Zhu et al. were able to convert an originally selected CD4 RNA aptamer into a DNA aptamer that is more stable, and that upon this conversion it exhibits a more potent inhibitory effect when conjugated to a siRNA against the HIV-1 protease, than the siRNA counterpart alone (Zhu et al. 2012).

As siRNA therapeutics is an area of immense interest and with several potentials in human disease, several other examples of siRNA delivery by means of cell-specific aptamers are expected to be published in the immediate future. It is the authors' opinion that given the increasing need for carrying large RNA payloads and for having multifunctional roles, aptamer development will be driven towards more specialized assemblies with opportunities for multiplexing, such as the more recent origami nanostructures (Table 1) (Soldevilla et al. 2018; Sakai et al. 2018).

Aptamer- miRNA Delivery

MicroRNAs (miRNAs) are small, endogenously expressed non-coding RNAs of 21-25 nucleotides long (Bader et al. 2011). They are master regulators of gene expression at the transcriptional and/or post-transcriptional level. They are encoded from nuclear DNA and their role is to target specific mRNAs for degradation or repression of translation. The biogenesis of miRNAs is a complex procedure. The genes encoding miRNAs are first transcribed to their corresponding primary miRNA (pri-miRNA) by the RNA Polymerase II, during which stage they acquire the characteristic stem-loop structure of pri-miRNAs. The conversion into precursor miRNA (pre-miRNA) requires cleavage of the 5' and 3' end of the pri-miRNA and this process is achieved by Drosha and DGCR8. With the help of Exportin 5, the pre-miRNA is transported out of the nucleus and once in the cytoplasm it is loaded onto Dicer that cleaves the loop generating the mature miRNA duplex. Finally, the duplex is loaded onto RNA induced silencing complex (RISC) and further processing by Argonaute removes and degrades the passenger (antisense) strand. The guide (sense) strand, forms the mature miRNA that binds to 3' untranslated region of the target mRNA.

Therapeutic miRNAs function similarly to endogenous miRNAs, although their processing varies slightly (Oliveto et al. 2017). They are introduced into the cytoplasm by means of conjugation (e.g., viruses, cationic lipids, polymers) in the form of pre-miRNAs or mature miRNA duplexes (Baumann and Winkler 2014). It is then recognized by the RISC complex onto which they are loaded and then processed as an endogenous miRNA (Figure 2A). Depending on how the endogenous miRNA is implicated in a disease, miRNA-based therapeutics fall either under the category of miRNA mimics or miRNA antagonists. The role of miRNA mimics is to restore endogenous miRNAs that show loss of function due to the disease, for example tumour suppressor genes. This is typically achieved via the delivery of artificial miRNAs to diseased tissues. On the contrary, miRNA antagonists inhibit endogenous miRNAs that acquire a gain-of-function role in diseases. Antagonists or antimiRs, as they are most often called, suppress their target in a similar manner like siRNAs or AONs (described later on) via the synthesis of an artificial, single stranded oligonucleotide with sequence complementarity to the endogenous miRNA. In addition, due to their single stranded nature and their mode of function, antimiRs are sometimes referred to as a subtype of AONs (Soldevilla et al. 2018).

MicroRNAs also guide RISC to mRNAs. Usually only part of a miRNA, known as the seed, pairs with the target mRNA. This imprecise matching allows miRNAs to target hundreds of endogenous mRNAs. In several instances, miRNAs have been implicated in the regulation of cancer-related genes and downstream cellular pathways and processes (Bader et al. 2011). These miRNAs may act as oncogenes (also known as oncomiRs) or tumour suppressors. Therefore, they have been long considered as potential therapeutic agents for the treatment of cancer. As miRNAs are expressed aberrantly and have a tendency for multiple transcript targets, their use as therapeutic agents implies the necessity for controlled deliver to affected cells. Following the line of siRNAs and the successful development of aptamer-siRNA chimeras for cell-specific delivery, aptamers have been recently explored as potential targeting moieties for cancer-related microRNAs as well (Soldevilla et al. 2018).

In one of the earlier studies, the well-known prostate specific membrane antigen (PSMA) RNA aptamer was explored for its role as a miRNA carrier molecule as well (X. Wu et al. 2011). In this particular study by Wu et al. the second generation A10-3.2 anti-PSMA aptamer was investigated as a targeting ligand for miR-15a and miR-16-1 tumour suppressors in prostate cancer cells. A positively charged spherical polymer, PAMAM (polyamidoamine) was first loaded with the therapeutic miRNAs and then conjugated through a polyethylene glycol (PEG) spacer to the aptamer. This construct was delivered with high specificity to LNCaP prostate cancer cells overexpressing PSMA, resulting in selective cell death.

In 2012, the Chen group published two independent studies reporting aptamer-miRNA directed apoptosis of the ovarian cancer cell line OVCAR3 via targeting of two different tumour suppressor miRNAs (N. Liu, Zhou, et al. 2012; Dai et al. 2012). In the first study, the let-7i miRNA was covalently conjugated to MUCIN 1 (MUC1) aptamer, an aptamer that preferentially binds to tumours expressing mucin 1 (Figure 3A, miRNA) (N. Liu, Zhou, et al. 2012). Chemoresistance in ovarian cancers has been associated with downregulation of the tumour suppressor let-7i. The results showed specific delivery of let-7i into tumour cells that subsequently reversed the resistance of OVCAR3 cells to paclitaxel, a chemotherapeutic drug, leading to inhibition of cell proliferation, induction of cell apoptosis, and decrease of cell survival *in vitro*. In the second study, Chen and colleagues induced cell apoptosis by conjugating the MUC1 aptamer to miR-29b (Dai et al. 2012). The specific delivery of miR-29b supressed the expression of DNA methyltransferesases that subsequently upregulated the expression of the PTEN gene. The product of the latter is an enzyme that acts as a tumour suppressor leading to apoptosis of ovarian carcinoma cells.

The GL21.T RNA aptamer developed against the Axl receptor tyrosine kinase (RTK) has been a popular aptamer for the delivery of several tumour-related miRNAs (Esposito et al. 2014; Russo et al. 2018; Esposito et al. 2016; Iaboni et al. 2016). Axl is an oncogenic RTK with key roles in tumorigenesis and metastasis of many cancers (R. Liu et al. 2010). In a study by Esposito et al. in 2014, the GL21.T aptamers was demonstrated to selectively delivery let-7g miRNA to A549 adenocarcinoma human alveolar basal epithelial cells (Axl positive cells) (Esposito et al. 2014). Furthermore, the conjugation of the tumour suppressor, let-7g miRNA, to the Axl aptamer which has a dual role (as a targeting moiety for the miRNA and antagonist for the Axl receptor) was able to induce inhibition of cell migration and cell survival *in vitro* and inhibition of tumour growth *in vivo*. The same aptamer was also conjugated to another tumour suppressor, miR-34c, for selective delivery to non-small cell lung cancer cells (NSCLC) with a subsequent effect on cell proliferation (Russo et al. 2018). The dual function of this aptamer was further utilized in the delivery of miR-137 tumour suppressor in glioblastoma stem-like cells (GSCs) (Esposito et al. 2016). GSCs have been implicated in the relapse and resistance of glioblastoma to therapeutic intervention. Therefore, the selective targeting of GSCs can potentially offer an effective treatment for glioblastoma. The GL21.T-miR-137 chimera was delivered along the Gint.4 RNA aptamer conjugated to anti-miR-10b (antagonists of the oncogenic miR-10b). Similarly to GL21.T aptamer, Gint.4 is a targeting moiety and an antagonist, selected against another RTK receptor, platelet derived growth factor β (PDGFβ). Combinatorial delivery of the aptamer-miRNA/anti-miR conjugates to the glioblastoma stem cell population resulted in inhibition of GSC propagation.

In 2016, Iaboni et al. demonstrated another application for the GL21.T aptamer (Iaboni et al. 2016). When conjugated to miR-212, the aptamer-miRNA chimera can be used as an adjuvant to TNF-related apoptosis-inducing ligand (TRAIL) therapy for the treatment of lung cancer. More specifically, the increase in the expression of the tumour suppressor miR-212, inhibited the anti-apoptotic protein PED/PEA-15 leading to TRAIL-mediate cytotoxicity to cancer cells.

The transferrin receptor which is universally expressed in all cells, in order to facilitate the import of iron, has also been explored as a potential delivery system for miR-126 (Rohde et al. 2015). In breast cancer in particular, the downregulation of miR-126 has implications in tumour growth, angiogenesis and metastasis. Therefore, this study sought to evaluate the ability of the TfR aptamer to selectively delivery miR-126 in breast cancer cells. As a second aim they compared the delivery and functionality of the mature miR-126-TfR aptamer as opposed to that of the pre-miR-126-TfR aptamer. The results showed that only the latter was able to generate functionally active miR-126, leading to increased levels of both miR-126 form (miR-126-3p and miR-126-5p) with subsequent reduction in tumour cell proliferation.

With the aim to restore radiotherapy-induced myelosuppression, the group of Zheng et al. explored the ability of c-kit aptamer to selectively delivery miR-26a mimic to heamatopoietic stem/ progenitor cells (HPSCs) (Tanno et al. 2017). The effect of the chimera was tested in HPSCs that highly express the c-kit receptor. The result here demonstrated the ability of miR-26a to protect haematopoiesis from chemotherapeutic agent-induced myelosuppression by targeting Bak1 (a pro apoptotic gene). These results were further evaluated on a mouse model of breast cancer. The miR-26a chimera was able to suppress tumour growth and to exert myeloprotection against a chemotherapeutic drug (5'fluorouracil). As miR-26a has been observed in other cancer types and the c-kit receptor is expressed in other forms of cancer, including stromal, small-cell lung cancer, glioblastoma and many more, the therapeutic utility of the miR-26a chimera could extend beyond breast cancer (Tanno et al. 2017).

Similarly, the novel aptamiR composed of the nucleolin aptamer AS1411 and the antimiR-21 (miRNA inhibitor, explained in antisense oligonucleotides section) chimera could have a therapeutic potential for various types of cancers (Pofahl, Wengel, and Mayer 2014). The antimiR domain was chemically modified with a phosphorothioate-LNA and covalently conjugated to the aptamer. Upon selective delivery in non-small-cell lung cancer cells A549 the antimiR targeted and supressed the endogenous miRNA 21 while the AS1411 exerted its anti-proliferative activity leading to prolonged inhibition of cancer cell growth. As both the aptamer and the therapeutic miRNA are expressed in many cancer types, the therapeutic potential of this aptamer-miRNA chimera is high.

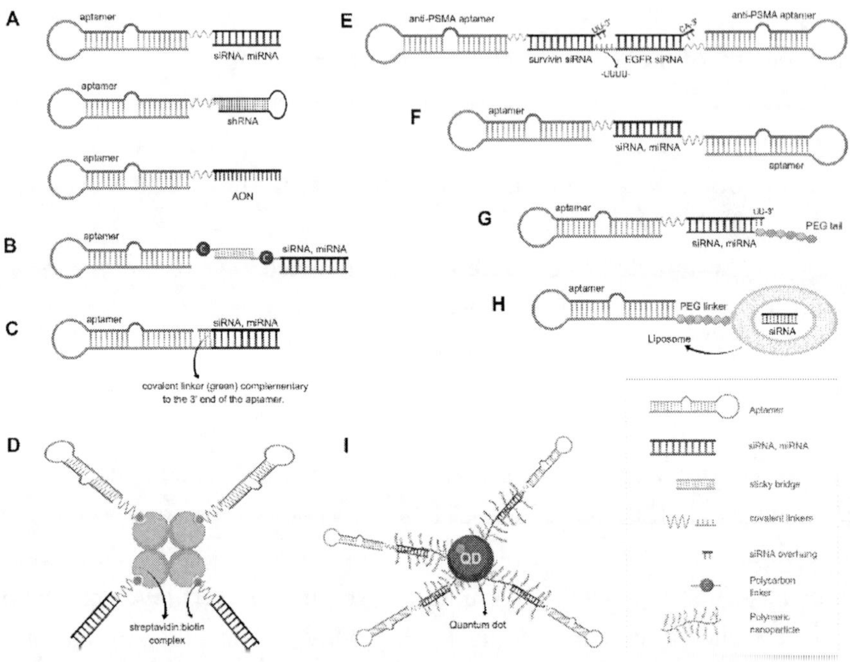

Figure 3. Aptamer-RNAi conjugation approaches. Schematics of common conjugation methods for RNAi sequences. A. Aptamer-siRNA, miRNA or AON chimeras conjugated via covalent linker, synthesized as a single molecule or as two parts that are later on annealed together. B. Aptamer-sticky bridge-siRNA, miRNA chimera. A 17nt, chemically modified, GC rich sequence serves as a linker to connect the aptamer with the therapeutic sequence. This approach serves as a universal linker for the conjugation of various siRNA, miRNAs. Furthermore, the sticky bridge is attached to the aptamer via a polycarbon linker. C. Aptamer-sticky bridge "like" approach. The 5' of the siRNA or miRNA is extended (green) with a sequence complementary to the 3' end of the aptamer (sequence shown in blue, complementary to the sequence shown in green) creating in this way a sticky bridge-like conjugation. D. Aptamer-streptavidin-siRNA complex. The aptamer and siRNAs are chemically modified with a biotin group which then facilitates binding to a streptavidin core. To facilitate release of the siRNA from the complex and subsequent RNAi processing, disulfide bonds are incorporated between the siRNA and the biotin that are reduced once in the acidic intracellular environment of the cell, resulting into siRNA release. E. Bivalent aptamer-siRNA chimera. Each aptamer is covalently conjugated to a siRNA and then the two parts are joined together via an additional covalent linker composed of a stretch of Uracyl bases (-UUUU-). In this way, two oncogenes can be targeted for silencing at the same time. Furthermore, the incorporation of two anti-PSMA aptamers rather than one, increased the specificity of the construct. Base modifications at the 3' end of each siRNA (-UU, -CA) were also added. F. Aptamer-siRNA chimera with increased specificity. Two aptamers targeting the same or different cell receptor are covalently linked with each

other via the therapeutic siRNA, miRNA. G. Second generation aptamer-siRNA or miRNA chimeras. The aptamer was truncated to its minimal functional size. A two nucleotide (-UU) overhang was added to the siRNA to enhance loading of the siRNA guide strand onto RISC. A polyethylene glycol (PEG) molecule was linked to the passenger strand for improved pharmacokinetic (PK) properties. H. Non-covalent aptamer-siRNA, miRNA conjugation via PEG. The PEG molecule serves as a linker and a spacer, to attach the aptamer to the therapeutic siRNA sequence while maintaining space for the aptamer portion to obtain its proper 3D conformation. At the same time, the PEG molecule improves pharmacokinetics and reduces renal clearance. The liposomal encapsulation enhances internalization of the siRNA. I. Example of an aptamer-functionalized nanoconstruct. The aptamer is non-covalently conjugated to a nanoarrier composed of a polymeric portion (polyethyleneimine, PEI) and a quantum dot (QD). The siRNA is encapsulated in PEI and covalently linked to the aptamer. This construct is then conjugated to the QD via electrostatic interactions creating PEI-coated QDs.

In another study from Catuogno et al., an antisense miRNA termed antagomiR were directed against tumour cells using two different aptamers as carrier molecules (Catuogno et al. 2015). AntimiRs or antagomirs are RNA-based antagonists for endogenous miRNAs and as therapeutics they are specifically targeted against disease-associated miRNAs (Stenvang et al. 2012). They silence/inhibit the function of specific miRNAs, by binding onto them thus preventing them from finding their mRNA target. The selected aptamers, Axl and platelet derived growth factor β (PDFRFβ) both bind and antagonize receptor tyrosine kinases that are cancer-related. Both aptamers were conjugated to the tumour suppressor anti-miR-222 and upon introduction in a human glioma cell line overexpressing either the Axl or the PDGFRβ receptor, they demonstrated selective and specific aptamer-mediated delivery of the anti-miR-222 (Catuogno et al. 2015). Furthermore, the selective delivery of the antagomiR was able to decrease miR-222 and increase the miR-222 target proteins. Additionally, this study set out to evaluate the effect of having two different antimiR sequences next to each other connected to a single aptamer. The results showed it was possible to simultaneously inhibit two different miRNAs without negatively affecting the specificity of the aptamer. This was also the first report of aptamer-mediated delivery of antimiRs suggesting the applicability of this approach for other diseases with similar pathologies.

Aptamer- shRNA Delivery

A shRNA is a siRNA with a hairpin structure (Figure 2B) (Moore et al. 2010). It consists of a sense and an antisense sequence of paired nucleotides that are separated by a loop sequence of unpaired nucleotides. shRNAs are typically introduced into cells in expression vectors (viral or bacterial) and once in the nucleus, they are processed by Drosha and then exported into the cytoplasm by Exportin-5 (Moore et al. 2010). In the cytoplasm, the shRNA associates with DICER, resulting in removal of the loop sequence and subsequent conversion into a siRNA. From this point onwards it is regarded and processed as a siRNA where degradation of the mRNA target is induced by the RNAi machinery.

The use of viral vectors for the delivery and stable expression of shRNA, while being advantageous for transfecting various cell types, at the same time, it poses safety risks which could severely limit their use as carrier molecules. These relate to potential off-target cell modifications by the shRNA as well as the risk of an immune response against the viral vector. In this regard, the efficacy of shRNA delivery can be improved with conjugation to aptamers that bind to cell surface receptors and internalize in target cells.

While the examples with their siRNA siblings are numerous, examples of aptamer-mediated shRNA delivery are limited. One possible explanation could be the finding of cell toxicity caused by some shRNA expression vectors, especially when administered *in vivo*, offering a therapeutic advantage to the exploration of siRNAs for aptamer guided delivery (McBride et al. 2008; Moore et al. 2010). Nevertheless, the following examples demonstrate the therapeutic utility of aptamer-shRNA constructs. The work of Chung-Il and Yokobayashi in 2006, was the first to demonstrate the temporal control of RNAi with an aptamer-shRNA chimera (An, Trinh, and Yokobayashi 2006). In this study, a theophylline selected aptamer was incorporated in the loop region of the shRNA designed to silence fluorescent reporter genes (EGFP and DsRed). Once administered in HEK293 cells aptamer binding to theophylline inhibited siRNA production of the aptamer-focused shRNA targeting the DsRed gene with a similar effect observed *in vivo*. The effect of the theophylline on the RNAi machinery was reversed with the removal of theophylline demonstrating a novel way to modulate RNAi processes through small molecules, without engineered proteins (An, Trinh, and Yokobayashi 2006).

Aptamer-shRNAs delivery has been investigated primarily in cancer therapy, in an attempt to selectively destruct cancer cells. Kim et al. showed an elegant way to selectively induce a high cell death effect of prostate cancer cells by utilizing co-delivery of a shRNA against Bcl-xL (an anti-apoptotic gene) and Doxorubicin (DOX) using aptamer-decorated polyplexes (E. Kim et al. 2010). The polyplexes were generated by conjugating branched polyethyleneimine (PEI) and hetero bifunctional polyethylene glycol (PEG) molecules as the shRNA delivery system. The anti-PSMA aptamer which is present on the surface prostate cancer cells was then conjugated to PEG and subsequently DOX was loaded onto the aptamer. DOX which is an effective anticancer drug, was intercalated into the nucleic acid bases of the aptamer's loop and lastly the Bcl-xL shRNA was encapsulated in the PEI-PEG-PSMA-DOX complexes. As demonstrated by the results of this study, the aptamer decorated complexes were able to selectively induce cell death in LNCaP cells (PSMA positive) and to further inhibit the proliferation of this prostate cancer cell line (E. Kim et al. 2010). In a similar manner, Askarian et al. published in 2015 another paradigm of aptamer-shRNA PEI polyplex for targeted delivery in lung cancer cells (Askarian et al. 2015). The Bcl-xL shRNA-encoding plasmid was conjugated to the nucleolin aptamer (AS1411) aptamer assisting its targeted delivery to A549 lung cancer cells but not into the negative control L929 cell line. The platform was composed of a poly-L-lysine (PLL) polymer core onto which the PEI polymer was conjugated via alkyl modification. Subsequently the aptamer and the shRNA, both negatively charged molecules, were electrostatically coupled to the positively charged polyplexes. The results of this work demonstrated efficient delivery of shRNA against Bcl-xL, with subsequent downregulation of the gene and selective induction of apoptosis in A549 cells.

Another potential therapeutic agent for prostate cancer was demonstrated by the work of Ni et al. (X. Ni, Zhang, et al. 2011). This group followed a different yet equally effective approach as Kim et al., for destructing cancer cells (E. Kim et al. 2010). Cell death in LNCaP prostate cancer cells was induced by a DNA-activated protein kinase (DNAPK) shRNA selectively delivered in these cells by the PSMA-targeting RNA aptamer (Figure 3A, shRNA). As the DNAPK is a key factor in DNA repair pathway, Ni et al. hypothesized that its inhibition via a shRNA, during ironizing irradiation (IR), can significantly enhance tumour response (X. Ni, Zhang, et al. 2011). The efficacy of the Bcl-xL shRNA-PSMA aptamer delivery system was assessed also in PSMA negative cells (PC3 cell line) as control. The results showed a selective reduction of DNAPK in LNCaP cells and also in LNCaP mouse xenografts. Subsequently, both were shown to be sensitized to IR therapy resulting to increased tumour cell death. Therefore this approach can be a promising tool for radiation therapy in prostate cancer.

Bcl-Xl shRNA delivery in lung cancer cells was achieved in a more recent study using a different nanocarrier, an alkyl modified PAMAM dendrimer conjugated to aptamer nucleolin (Ayatollahi et al. 2017). In this work the polyamidoamine (PAMAM) dendrimer was loaded with shRNA and subsequently conjugated to the aptamer via covalent (amine modification) or noncovalent conjugation (electrostatic). The delivery system was assessed *in vitro* in A549 lung cancer cells where significant decrease in Bcl-xL expression was shown with subsequent selective induction of apoptosis, as no cell death was observed in the negative cell line, L929. As the authors indicated, the development of the AS1411-PAMAM-shRNA nanoparticle could be used for gene therapy applications for pulmonary diseases (Ayatollahi et al. 2017).

The work by Pang et al. demonstrates the application of aptamer-shRNA chimeras for a wider range of diseases, including infectious diseases such as Human Immunodeficiency Virus type 1 infection (HIV-1) (Pang et al. 2018). Among other constructs, the authors designed a chimera made of a newly developed aptamer against HIV-1 integrase (aptamer S_3R_3) which is required for the integration of viral DNA in the host genome, and a shRNA targeting the tat-rev region. The fusion of these two resulted in a very strong and prolong inhibition of HIV replication in cell cultures that could be useful in the future for gene therapy applications to confer HIV resistance and potentially a cure for AIDS (acquired immunodeficiency syndrome).

Aptamer - AON Delivery

Antisense oligonucleotides (AONs) function similarly to a miRNA antagonist. AONs are a class of synthetic, single stranded RNA molecules that have sequence complementarity to their mRNA target, thus the term "antisense" (Figure 2C) (Dias and Stein 2002). Their size ranges between 18 and 21 nucleotides and are usually chemically modified to enhance their stability and nuclease resistance against serum nucleases. AONs are designed to specifically hybridize to a unique mRNA target sequence through Watson-Crick base pairing, modulating in this way gene expression at the mRNA level (Dias and Stein 2002). This is a highly promising therapeutic strategy for several monogenic diseases, including thalassaemia, muscular dystrophies and several neurological disorders (Wurster and Ludolph 2018; El-Beshlawy et al. 2008; Relizani et al. 2017; van Putten et al. 2019; Jauvin et al. 2017).

AONs function through different molecular mechanisms (Shen and Corey 2018). However, the majority of AON drugs that are currently investigated, both clinically and preclinical, function via RNAse H dependent mechanism that induce mRNA degradation or though splicing modulation of the precursor mRNA leading to an exon to be included or exploded from the mRNA thus altering protein translation (Dias and Stein 2002).

Similarly to siRNAs, shRNAs and microRNAs, the major challenge in the effective use of AONs is the specific delivery of AONs to their intracellular targets (Nguyen and Yokota 2019). The ability of aptamers to bind with high affinity and specificity to target, suggests their effectiveness in this category of RNAi effector molecules as well, as targeting moieties for cell-specific delivery. In a proof-of-principle study, Kotula et al.

demonstrated AON delivery by means of a cell targeting aptamer (Figure 3A, AON) (Kotula et al. 2012). In this study, the group utilized the AS1411 aptamer against nucleolin, a receptor known to be present on the cell surface of many types of cancer cells, offering a therapeutic potential for various forms of cancer. A key observation of this study was the ability of the aptamer to internalize and localize to the nucleus of cancer cells. With this in mind, the authors aimed to investigate the ability of the AS1411 aptamer to deliver a splice-switching AON to the nucleus. To assess the delivery of the AON, the aptamer-AON chimera was transfected in the prostate cancer cell line PC3 containing a luciferase reported construct with a premature stop codon. Successful delivery of the AON to the nucleus was able to alter slicing and induce luciferace production. This result is a strong example of the potential of AONs to alter mRNA splicing patterns as a means of therapy.

AONs are promising therapeutics for a number of monogenic diseases, including muscular dystrophies. More specifically, our group recently described in a proof-of concept-study the development of a skeletal muscle internalizing RNA aptamer (A01B) that has the potential to deliver therapeutic RNA sequences for use in Duchenne muscular dystrophy (DMD) (Philippou et al. 2018). DMD is caused by mutations in the dystrophin gene that disrupt the open reading frame (ORF) and result in the production of a dystrophin protein that is non-functional and unstable (Aartsma-Rus et al. 2017). Furthermore, studies have shown that internal deletions or duplications at the pre-mRNA level can restore the open reading frame and result in the production of dystrophin that is shorter but more stable, and functional. This approach converts the disease into a less severe form of muscular dystrophy, Becker muscular dystrophy and it can be mediated by AONs- an approach commonly referred to as AON-mediated exon skipping.

Our more recent investigations focus on extending this approach to animal models, in an attempt to create a more clinically relevant aptamer (unpublished data). In this attempt, the more recent aptamer selection method, *in vivo* SELEX, is employed and the therapeutic AON is incorporated in the selection process. This is expected to offer a therapeutic advantage to the selected aptamer given that the remains functional.

Aptamer-mediated delivery of AONs is a work still in progress. Nevertheless, the potential of aptamers to guide AONs to desired target(s) in the body is evident and there is excitement to what the future holds for these therapeutic approaches.

APTAMER-MEDIATED DELIVERY: CHALLENGES AND SOLUTIONS

Nuclease Degradation

One of the first challenges in aptamer-based therapeutics is nuclease degradation (Dutta et al. 2016). Due to nuclease degradation, unmodified aptamers have an estimated circulation time of only 10 minutes. Therefore, in addition to the chemical substitution on the 2'position on the sugar ring of nucleotides (2'F, 2'NH2, and 2'OMe) upon their identification, aptamers can be further modified for preclinical and clinical applications. As such, additional post-SELEX modification sites include the ends of the oligonucleotide sequence, bases and the phosphate groups of the phosphodiester linkage. Capping of the 3'end with an inverted thymidine or an inverted biotin is a common chemical modification that increases nuclease resistance against 3' exonulcelases and enhances binding affinity for aptamers in clinical trials(Esposito et al. 2011). Examples of aptamers in clinical trials that accommodate a 3'end inverted thymidine, include Macugen, ARC1905 and BAX499 (Dutta et al. 2016). Moreover, aptamers that are intended for clinical applications, such as the anti-von Willebrand factor aptamer *ARC1779*, could be further protected against nucleases by adding changes to the phosphodiester backbone such as a phosphorothioate (PS) or boranophosphate (Hirano, Munnik, and Sato 2015). More recent chemical modifications such as the LNA technology and the spiegelmers that are "mirror" RNA sequences, results in structural changes in the aptamer sequence that cannot be recognized by nucleases (Y.M. Kim et al. 2015; Solinger and Spang 2014). On the other hand, the ribonucleoside analogue UNA (unlocked nucleic acid) can enhance the flexibility of the aptamer, as shown in a published work with the thrombin aptamer (Wavre-Shapton et al. 2014).

As the aptamer recognizes its target based on its structure, any post-SELEX modifications risk altering the folding pattern and subsequent function of the aptamer. As each aptamer is unique, no regulations apply modifications, but care should be taken to maintain the binding domains intact.

Renal Filtration

Renal filtration is the second challenge in aptamer-based therapeutics (Dutta et al. 2016; J. Zhou and Rossi 2016). Aptamers have a relatively small size, typically between 6-30 kDA (<5 nm in diameter) that is right below the renal glomerulus cut-off (between 30-50 kDA). As a consequence of this, when the aptamers are administered in the bloodstream they are rapidly excrete via the renal filtration leading to a very short half-life. To overcome the size limitation, bulky groups can be conjugated at the 5'-end of aptamers, such as high molecular weight PEG (usually between 20-40 kDA), cholesterol, proteins, liposomes, organic and inorganic nanomaterials (G.H. Kim et al. 2014; Scott, Vacca, and Gruenberg 2014; Puchner et al. 2013; Hubner and Peter 2012; Chinen et al. 2013). What these molecules do, is increase the overall size of aptamers above the cut-off limit of the kidneys. This in turn, enhances the pharmacokinetic profile of aptamers by extending their circulation times in serum and preventing or slowing down their renal clearance. Moreover, some of these molecules such as cholesterol, depending on the protein they are conjugated to, can also facilitate delivery to specific organs, such as the liver. PEG is an FDA-approved formulation, known to prolong circulation half-life and improve the *in vivo* bioavailability of aptamers upon systemic delivery. Conjugation of PEG to the RNA aptamer Macugen, increased its serum stability from 9 to 12 hours following systemic administration (Beas et al. 2012). While, PEG is a well-studied formulation, immune response against PEG was observed in some patients of a phase III clinical trial of the aptamer-based anticoagulant system, REG1 (carries an RNA aptamer and an antidote oligonucleotide) (Griffiths et al. 2012). It was later on found that the presence of pre-existing PEG antibodies was responsible for this acute immune response.

Intracellular Fate of Aptamers

A third challenge for all nucleic acid therapeutics, not limited to aptamers, is the cellular fate of nucleic acids following internalization. Currently, several aptamer-drug complexes have exploited internalization; however, only a few studies further explored the mode of cellular delivery (Orava, Cicmil, and Gariepy 2010). There are several possible routes of internalization, including receptor turnover, membrane recycling, endosomal internalization and endocytosis (McNamara et al. 2006b; Benedetto et al. 2015; J. Zhou et al. 2009b; J. Zhou et al. 2008a; Dassie et al. 2009a; Tawiah, Porciani, and Burke 2017). The findings of most of these examples suggest entry via receptor-mediated endocytosis and subsequent localization in endocytic vesicles such as endosomes However, whether internalization occurs via aptamer binding on a receptor or due to natural membrane turnover is yet unknown (J. Zhou et al. 2009b). What it is known and used for an initial characterization of aptamer's mode of entry, is that at cold temperatures (4°C), the membrane turnover is compromised and thus the aptamer is forced to bind on the surface (Daniels et al. 2003). Increasing the temperature back to 37°C, induces aptamer internalization most likely through the receptor onto which the aptamer initially bound (Serfass et al. 2017; W.H. Thiel, Bair, et al. 2012; Y. Wang, Luo, et al. 2014). In another study using epithelial ovarian carcinoma cells, the mode of endocytosis and subcellular localization of the identified aptamer was monitored by confocal imaging (Benedetto et al. 2015). Using an endosomal-specific marker (pHrodo Red Transferrin: bright fluorescence when localized in endocytic compartments), the authors obtained images that supported colocalization of the aptamer in endosomal structures. These data suggested that the internalization of this aptamer is regulated by endocytic pathways that almost always direct the cargo to early endosomes.

Unless the aptamer finds a way to escape the early endosome it will be targeted for degradation (Tawiah, Porciani, and Burke 2017). This occurs through maturation of early endosomes to late endosomes and subsequent fusion with lysosomes. Notably, studies with aptamer-siRNAs (small interfering RNAs) conjugates have been shown to escape the endosome (McNamara et al. 2006b; Benedetto et al. 2015; J. Zhou et al. 2009b; J. Zhou et al. 2008a; Dassie et al. 2009a; Tawiah, Porciani, and Burke 2017). This suggestion was based, initially, on the observation that siRNAs must localize to the cytoplasmic compartment of a cell to exert their function. Other studies, led, later on, to the suggestion that at least the therapeutic siRNA is dissociated from the aptamer and escapes the endosomes (Neff et al. 2011a; McNamara et al. 2006b; J. Zhou et al. 2008a; K.W. Thiel, Hernandez, et al. 2012a). A more recent, however, publication by the Giangrande group, reported a functional assay (RIP assay) to confirm cellular uptake and subsequent cytoplasmic release of RNA aptamers that bind to cell surface receptors (Law et al. 2017). In this study, a cellular toxin (saporin) with cytotoxic properties was conjugated to the A9g aptamer (without any therapeutic oligonucleotide). Upon aptamer-mediated delivery of saporin to the cytoplasm, the toxin exerted its ribosome inactivating protein effects, thus leading to cell death. While this is an important advancement for cell-internalizing aptamers that are developed for RNAi delivery (siRNAs, and microRNAs), the exact mechanism of endosomal escape remains yet unclear.

Toxicity

Currently, there is limited knowledge about the toxicity of aptamers in clinical trials (T.T. Liu, Gomez, et al. 2012; Takahashi et al. 2012). While, minimal side effects have been reported, the anionic nature of aptamers and the use of chemical modification or conjugations could potentially cause toxicity, particularly with repeated administrations (Huotari et al. 2012; van

Weering, Verkade, and Cullen 2012; Huotari and Helenius 2011). For example, in the work by Swayze et al. LNA containing oligonucleotides induced hepatotoxicity as early the 4th day, after a single administration, indicating the potential risks of this chemistry (Kanwar and Fortini 2008). However, as indicated by another work with 2'F and 2'OMe containing nucleotides, the activation of the immune system by therapeutic RNAs can be advantageous for treating some diseases, such as cancers (known as cancer immunotherapy), as long as these immune and inflammatory responses can be controlled (Shah et al. 2007). As previously mentioned in the "Renal filtration" section, similar adverse effects can be obtained from conjugations with high molecular weight formulations, such as the PEG group. Additionally, high lipophilic molecules, such as liposomes and cholesterol, have high potency for liver uptake than, can, in turn, cause hepatotoxicity (Murphy 1991). These formulations raise some safety issues about the long-term use of these agents in humans that will have to be improved before aptamers can truly enter the clinic.

A more recent strategy that could potentially address all four challenges, is the rational multimerization and/or multiplexing of aptamers (Wilson et al. 2000; Szeto et al. 2013; Economou et al. 2017). This strategy, creates a new multivalent molecule with a molecular weight above the threshold of the kidneys. These newer generation of aptamers, have shown improved performance compared with single aptamers, including higher binding affinity and specificity, biological function and as expected, circulation time. To date, such promising results have been obtained with the dimerization (2 molecules) and tetramerization (4 molecules) of single aptamers (Wallner et al. 2017).

FUNCTIONALIZED NANOPARTICLES: THE FUTURE OF APTAMER-MEDIATED DELIVERY

Following the line of multimerization, aptamer-conjugated nanoparticles are vastly becoming the trend in aptamer-mediated delivery, as discussed earlier, for the delivery of RNAi effector molecules. This is due to their ability to codeliver multiple components, such as a chemotherapeutic agents, imaging agents, RNAi effector molecules, and (an) aptamer(s), at once (Figure 4).

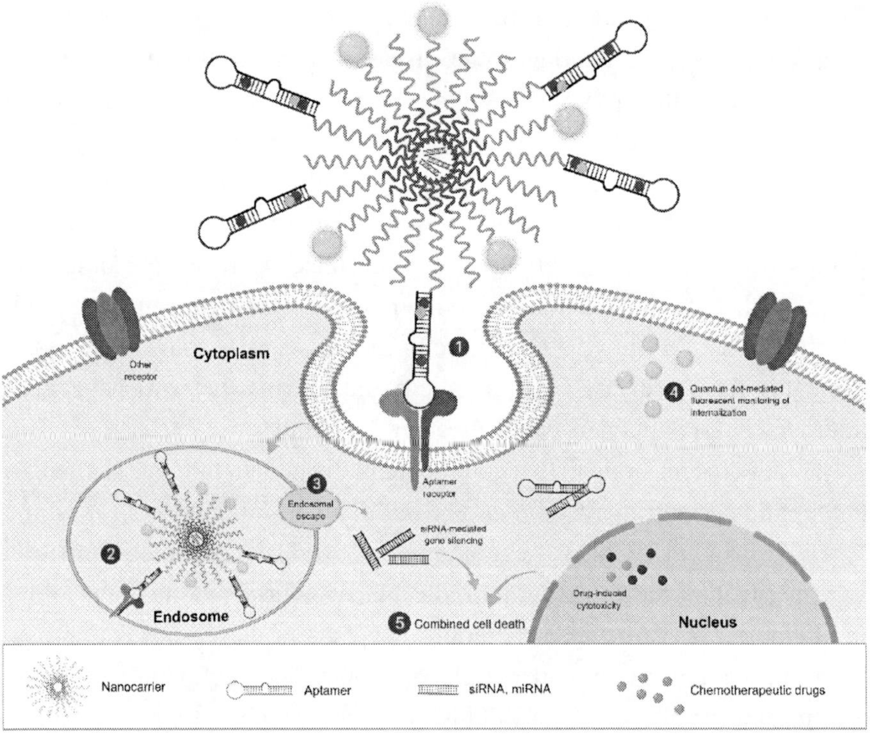

Figure 4. Aptamer-functionalized nanoparticle. Schematic demonstrating the potential therapeutic applications of aptamers as cell targeting moieties for nanoparticles. Nanoparticles can be loaded with multiple groups including therapeutic sequences (siRNA), fluorescent agents (quantum dots) for intracellular monitoring and/or drugs such as chemotherapeutic drugs. Multiplexing results into the creation of multifunctional nanoparticles with numerous applications, including the combined death of cancerous cells as demonstrated above.

Nanoparticles are structures which belong in the nanometre range, usually less than 100 mm. These structures should be able to pass through vessels that are accumulating in the affected areas. However, a key difference of nanoparticles from other molecules is that they have a large surface area. Therefore, they can achieve large payloads for diagnostic and therapeutic agent applications (M. Kim et al. 2018; Sivakumar et al. 2019). Molecules like aptamers, can be attached onto the surfaces of nanoparticles to ensure high specificity. This strategy is referred to as active targeting.

Aptamer-conjugated nanostructures include compounds like gold nanoparticles, oxide nanoparticles, dendrimers, carbon nanotubes, liposomes, micelles, quantum dots and polymers (Carolina de Agular Ferreira 2013). As delivery systems, aptamer-conjugated nanoparticles have several advantages, including their ability to affect the biodistribution and pharmacokinetics of drugs, the increased permeability across the cell membrane, the ability to carry several and different cargos, and finally the ability to exert cell specific targeting owed to their conjugation with aptamers. Furthermore, aptamers have the potential of being chemically synthesized. Therefore, other functional groups can be added to their ends to facilitate their linkage to nanoparticles. Additionally, aptamers are also nanoscale molecules. Therefore, they have little-to-no effect on the overall size and dimension of the nano-drug. This is critical in the passive uptake and transportation of the nano-drug through the microvasculature.

After combining the advantages of nanoparticles and the tissue-targeting capabilities of aptamers, the use of aptamer-functionalized nanoparticles may be the awaited solution in the field of biomedicine (Carolina de Agular Ferreira 2013).

CONCLUSION

Since its inception, the use of RNAi technology has revolutionized how we perform research on gene function. Furthermore, the use of aptamers as cell-specific targeting moieties has positioned aptamer-RNAi conjugates as one of the most promising alternatives to traditional medication. Perhaps, the most important advantage is their three-dimensional structure that allows them to bind to specific targets on cells with high affinity and specificity. However, the use of aptamers in the field of biomedicine is likely to include further challenges, in addition to some exciting new applications.

In this chapter, an up-to-date review of the most relevant information on aptamer development and aptamer-targeted delivery of RNAi conjugates, including siRNA, microRNA, shRNA and AONs as well as aptamer challenges and solutions is provided. This review should be useful for researchers interested in engaging with the development of aptamers for delivery purposes as well as researchers looking for alternative carrier molecules with superior delivery characteristics, cell specificity in particular.

Acknowledgments

This work was supported by a grant to Professor Leonida A. Phylactou from the A.G. Leventis Foundation (Leventis Foundation).

References

Aartsma-Rus, A., V. Straub, R. Hemmings, M. Haas, G. Schlosser-Weber, V. Stoyanova-Beninska, E. Mercuri, F. Muntoni, B. Sepodes, E. Vroom, and P. Balabanov. 2017. "Development of Exon Skipping Therapies for Duchenne Muscular Dystrophy: A Critical Review and a Perspective on the Outstanding Issues." *Nucleic Acid Ther* 27 (5): 251-259. https://doi.org/10.1089/nat.2017.0682.

Altieri, D. C. 2013. "Targeting survivin in cancer." *Cancer Lett* 332 (2): 225-8. https://doi.org/10.1016/j.canlet.2012.03.005.

An, C. I., V. B. Trinh, and Y. Yokobayashi. 2006. "Artificial control of gene expression in mammalian cells by modulating RNA interference through aptamer-small molecule interaction." *RNA* 12 (5): 710-6. https://doi.org/10.1261/rna.2299306.

Askarian, S., K. Abnous, S. Taghavi, R. K. Oskuee, and M. Ramezani. 2015. "Cellular delivery of shRNA using aptamer-conjugated PLL-alkyl-PEI nanoparticles." *Colloids Surf B Biointerfaces* 136: 355-64. https://doi.org/10.1016/j.colsurfb.2015.09.023.

Ayatollahi, S., Z. Salmasi, M. Hashemi, S. Askarian, R. K. Oskuee, K. Abnous, and M. Ramezani. 2017. "Aptamer-targeted delivery of Bcl-xL shRNA using alkyl modified PAMAM dendrimers into lung cancer cells." *Int J Biochem Cell Biol* 92: 210-217. https://doi.org/10.1016/j.biocel.2017.10.005.

Bader, A. G., D. Brown, J. Stoudemire, and P. Lammers. 2011. "Developing therapeutic microRNAs for cancer." *Gene Ther* 18 (12): 1121-6. https://doi.org/10.1038/gt.2011.79.

Bagalkot, V., and X. Gao. 2011. "siRNA-aptamer chimeras on nanoparticles: preserving targeting functionality for effective gene silencing." *ACS Nano* 5 (10): 8131-9. https://doi.org/10.1021/nn202772p.

Baumann, V., and J. Winkler. 2014. "miRNA-based therapies: strategies and delivery platforms for oligonucleotide and non-oligonucleotide agents." *Future Med Chem* 6 (17): 1967-84. https://doi.org/10.4155/fmc.14.116.

Beas, A. O., V. Taupin, C. Teodorof, L. T. Nguyen, M. Garcia-Marcos, and M. G. Farquhar. 2012. "Galphas promotes EEA1 endosome maturation and shuts down proliferative signaling through interaction with GIV (Girdin)." *Mol Biol Cell* 23 (23): 4623-34. https://doi.org/10.1091/mbc.E12-02-0133.

Benedetto, G., T. J. Hamp, P. J. Wesselman, and C. Richardson. 2015. "Identification of epithelial ovarian tumor-specific aptamers." *Nucleic acid therapeutics* 25 (3): 162-72. https://doi.org/10.1089/nat.2014.0522.

Burmeister, P. E., S. D. Lewis, R. F. Silva, J. R. Preiss, L. R. Horwitz, P. S. Pendergrast, T. G. McCauley, J. C. Kurz, D. M. Epstein, C. Wilson, and A. D. Keefe. 2005. "Direct in vitro selection of a 2'-O-methyl aptamer to VEGF." *Chem Biol* 12 (1): 25-33. https://doi.org/10.1016/j.chembiol.2004.10.017.

Carolina de Agular Ferreira, Andre Luis Branco de Barros. 2013. "Aptamer Functionalized Nanoparticles for Cancer Targeting." *Molecular Pharmaceutics & Organic Process Research* 1 (2): 2. https://doi.org/10.4172/2329-9053.1000105.

Catuogno, S., A. Rienzo, A. Di Vito, C. L. Esposito, and V. de Franciscis. 2015. "Selective delivery of therapeutic single strand antimiRs by aptamer-based conjugates." *J Control Release* 210: 147-59. https://doi.org/10.1016/j.jconrel.2015.05.276.

Chen, K., B. Liu, B. Yu, W. Zhong, Y. Lu, J. Zhang, J. Liao, J. Liu, Y. Pu, L. Qiu, L. Zhang, H. Liu, and W. Tan. 2017. "Advances in the development of aptamer drug conjugates for targeted drug delivery." *Wiley Interdiscip Rev Nanomed Nanobiotechnol* 9 (3). https://doi.org/10.1002/wnan.1438.

Chen, L., F. Rashid, A. Shah, H. M. Awan, M. Wu, A. Liu, J. Wang, T. Zhu, Z. Luo, and G. Shan. 2015. "The isolation of an RNA aptamer targeting to p53 protein with single amino acid mutation." *Proceedings of the National Academy of Sciences of the United States of America* 112 (32): 10002-7. https://doi.org/10.1073/pnas.1502159112.

Chen, Y., K. Yu, J. B. Wu, J. L. Zhang, X. D. Hu, L. Jiang, S. H. Zhang, X. X. Hu, and S. M. Gao. 2004. "[The protective effect of amifostine on hydroquinone-induced apoptosis in bone marrow]." *Zhonghua Lao Dong Wei Sheng Zhi Ye Bing Za Zhi* 22 (3): 165-7. http://www.ncbi.nlm.nih.gov/pubmed/15256145.

Chen, Z., and X. Xu. 2016. "Roles of nucleolin. Focus on cancer and anti-cancer therapy." *Saudi Med J* 37 (12): 1312-1318. https://doi.org/10.15537/smj.2016.12.15972.

Cheng, C., Y. H. Chen, K. A. Lennox, M. A. Behlke, and B. L. Davidson. 2013. "In vivo SELEX for Identification of Brain-penetrating Aptamers." *Molecular therapy. Nucleic acids* 2: e67. https://doi.org/10.1038/mtna.2012.59.

Chinen, T., S. Kazami, Y. Nagumo, I. Hayakawa, A. Ikedo, M. Takagi, A. Yokosuka, N. Imamoto, Y. Mimaki, H. Kigoshi, H. Osada, and T. Usui. 2013. "Glaziovianin A prevents endosome maturation via inhibiting microtubule dynamics." *ACS chemical biology* 8 (5): 884-9. https://doi.org/10.1021/cb300641h.

Chu, T. C., K. Y. Twu, A. D. Ellington, and M. Levy. 2006. "Aptamer mediated siRNA delivery." *Nucleic Acids Res* 34 (10): e73. https://doi.org/10.1093/nar/gkl388.

Cory, S., and J. M. Adams. 2005. "Killing cancer cells by flipping the Bcl-2/Bax switch." *Cancer Cell* 8 (1): 5-6. https://doi.org/10.1016/j.ccr.2005.06.012.

Dai, F., Y. Zhang, X. Zhu, N. Shan, and Y. Chen. 2012. "Anticancer role of MUC1 aptamer-miR-29b chimera in epithelial ovarian carcinoma cells through regulation of PTEN methylation." *Target Oncol* 7 (4): 217-25. https://doi.org/10.1007/s11523-012-0236-7.

Daniels, D. A., H. Chen, B. J. Hicke, K. M. Swiderek, and L. Gold. 2003. "A tenascin-C aptamer identified by tumor cell SELEX: systematic evolution of ligands by exponential enrichment." *Proceedings of the National Academy of Sciences of the United States of America* 100 (26): 15416-21. https://doi.org/10.1073/pnas.2136683100.

Darmostuk, M., S. Rimpelova, H. Gbelcova, and T. Ruml. 2015. "Current approaches in SELEX: An update to aptamer selection technology." *Biotechnol Adv* 33 (6 Pt 2): 1141-61. https://doi.org/10.1016/j.biotechadv.2015.02.008.

Dassie, J. P., X. Y. Liu, G. S. Thomas, R. M. Whitaker, K. W. Thiel, K. R. Stockdale, D. K. Meyerholz, A. P. McCaffrey, J. O. McNamara, 2nd, and P. H. Giangrande. 2009a. "Systemic administration of optimized aptamer-siRNA chimeras promotes regression of PSMA-expressing tumors." *Nature biotechnology* 27 (9): 839-49. https://doi.org/10.1038/nbt.1560.

---. 2009b. "Systemic administration of optimized aptamer-siRNA chimeras promotes regression of PSMA-expressing tumors." *Nat Biotechnol* 27 (9): 839-49. https://doi.org/10.1038/nbt.1560.

Dias, N., and C. A. Stein. 2002. "Antisense oligonucleotides: basic concepts and mechanisms." *Mol Cancer Ther* 1 (5): 347-55. http://www.ncbi.nlm.nih.gov/pubmed/12489851.

Dutta, S., S. Roy, N. S. Polavaram, M. J. Stanton, H. Zhang, T. Bhola, P. Honscheid, T. M. Donohue, Jr., H. Band, S. K. Batra, M. H. Muders, and K. Datta. 2016. "Neuropilin-2 Regulates Endosome Maturation and EGFR Trafficking to Support Cancer Cell Pathobiology." *Cancer research* 76 (2): 418-28. https://doi.org/10.1158/0008-5472.CAN-15-1488.

Eckerdt, F., J. Yuan, and K. Strebhardt. 2005. "Polo-like kinases and oncogenesis." *Oncogene* 24 (2): 267-76. https://doi.org/10.1038/sj.onc.1208273.

Economou, M., O. Kolokotroni, I. Paphiti-Demetriou, C. Kouta, E. Lambrinou, E. Hadjigeorgiou, V. Hadjiona, F. Tryfonos, E. Philippou, and N. Middleton. 2017. "Prevalence of breast-feeding and exclusive breast-feeding at 48 h after birth and up to the sixth month in Cyprus: the BrEaST start in life project." *Public Health Nutr*: 1-14. https://doi.org/10.1017/S1368980017003214.

El-Beshlawy, A., A. Mostafa, I. Youssry, H. Gabr, I. M. Mansour, M. El-Tablawy, M. Aziz, and I. R. Hussein. 2008. "Correction of aberrant pre-mRNA splicing by antisense oligonucleotides in beta-thalassemia Egyptian patients with IVSI-110 mutation." *J Pediatr Hematol Oncol* 30 (4): 281-4. https://doi.org/10.1097/MPH.0b013e3181639afe.

Ellington, A. D., and J. W. Szostak. 1990. "In vitro selection of RNA molecules that bind specific ligands." *Nature* 346 (6287): 818-22. https://doi.org/10.1038/346818a0.

Esposito, C. L., S. Catuogno, V. de Franciscis, and L. Cerchia. 2011. "New insight into clinical development of nucleic acid aptamers." *Discovery medicine* 11 (61): 487-96. http://www.ncbi.nlm.nih.gov/pubmed/21712014.

Esposito, C. L., L. Cerchia, S. Catuogno, G. De Vita, J. P. Dassie, G. Santamaria, P. Swiderski, G. Condorelli, P. H. Giangrande, and V. de Franciscis. 2014. "Multifunctional aptamer-miRNA conjugates for targeted cancer therapy." *Mol Ther* 22 (6): 1151-1163. https://doi.org/10.1038/mt.2014.5.

Esposito, C. L., S. Nuzzo, S. A. Kumar, A. Rienzo, C. L. Lawrence, R. Pallini, L. Shaw, J. E. Alder, L. Ricci-Vitiani, S. Catuogno, and V. de Franciscis. 2016. "A combined microRNA-based targeted therapeutic approach to eradicate glioblastoma stem-like cells." *J Control Release* 238: 43-57. https://doi.org/10.1016/j.jconrel.2016.07.032.

Farokhzad, O. C., S. Jon, A. Khademhosseini, T. N. Tran, D. A. Lavan, and R. Langer. 2004. "Nanoparticle-aptamer bioconjugates: a new approach for targeting prostate cancer cells." *Cancer Res* 64 (21): 7668-72. https://doi.org/10.1158/0008-5472.CAN-04-2550.

Geiger, A., P. Burgstaller, H. von der Eltz, A. Roeder, and M. Famulok. 1996. "RNA aptamers that bind L-arginine with sub-micromolar dissociation constants and high enantioselectivity." *Nucleic acids research* 24 (6): 1029-36. http://www.ncbi.nlm.nih.gov/pubmed/8604334.

Germer, K., M. Leonard, and X. Zhang. 2013. "RNA aptamers and their therapeutic and diagnostic applications." *International journal of biochemistry and molecular biology* 4 (1): 27-40. http://www.ncbi.nlm.nih.gov/pubmed/23638319.

Godfrey, C., L. R. Desviat, B. Smedsrod, F. Pietri-Rouxel, M. A. Denti, P. Disterer, S. Lorain, G. Nogales-Gadea, V. Sardone, R. Anwar, S. El Andaloussi, T. Lehto, B. Khoo, C. Brolin, W. M. van Roon-Mom, A. Goyenvalle, A. Aartsma-Rus, and V. Arechavala-Gomeza. 2017. "Delivery is key: lessons learnt from developing splice-switching antisense therapies." *EMBO Mol Med* 9 (5): 545-557. https://doi.org/10.15252/emmm.201607199.

Griffiths, R. E., S. Kupzig, N. Cogan, T. J. Mankelow, V. M. Betin, K. Trakarnsanga, E. J. Massey, S. F. Parsons, D. J. Anstee, and J. D. Lane. 2012. "The ins and outs of human reticulocyte maturation: autophagy and the endosome/exosome pathway." *Autophagy* 8 (7): 1150-1. https://doi.org/10.4161/auto.20648.

Hall, B., S. Arshad, K. Seo, C. Bowman, M. Corley, S. D. Jhaveri, and A. D. Ellington. 2009. "In vitro selection of RNA aptamers to a protein target by filter immobilization." *Curr Protoc Mol Biol* Chapter 24: Unit 24 3. https://doi.org/10.1002/0471142727.mb2403s88.

Hernandez, L. I., K. S. Flenker, F. J. Hernandez, A. J. Klingelhutz, J. O. McNamara, 2nd, and P. H. Giangrande. 2013a. "Methods for Evaluating Cell-Specific, Cell-Internalizing RNA Aptamers." *Pharmaceuticals* 6 (3): 295-319. https://doi.org/10.3390/ph6030295.

---. 2013b. "Methods for Evaluating Cell-Specific, Cell-Internalizing RNA Aptamers." *Pharmaceuticals (Basel)* 6 (3): 295-319. https://doi.org/10.3390/ph6030295.

Hirano, T., T. Munnik, and M. H. Sato. 2015. "Phosphatidylinositol 3-Phosphate 5-Kinase, FAB1/PIKfyve Kinase Mediates Endosome Maturation to Establish Endosome-Cortical Microtubule Interaction in Arabidopsis." *Plant Physiol* 169 (3): 1961-74. https://doi.org/10.1104/pp.15.01368.

Hong, T. M., Y. L. Chen, Y. Y. Wu, A. Yuan, Y. C. Chao, Y. C. Chung, M. H. Wu, S. C. Yang, S. H. Pan, J. Y. Shih, W. K. Chan, and P. C. Yang. 2007. "Targeting neuropilin 1 as an antitumor strategy in lung cancer." *Clin Cancer Res* 13 (16): 4759-68. https://doi.org/10.1158/1078-0432.CCR-07-0001.

Howe, L. R., and P. H. Brown. 2011. "Targeting the HER/EGFR/ErbB family to prevent breast cancer." *Cancer Prev Res (Phila)* 4 (8): 1149-57. https://doi.org/10.1158/1940-6207.CAPR-11-0334.

Hubner, M., and M. Peter. 2012. "Cullin-3 and the endocytic system: New functions of ubiquitination for endosome maturation." *Cell Logist* 2 (3): 166-168. https://doi.org/10.4161/cl.20372.

Huotari, J., and A. Helenius. 2011. "Endosome maturation." *EMBO J* 30 (17): 3481-500. https://doi.org/10.1038/emboj.2011.286.

Huotari, J., N. Meyer-Schaller, M. Hubner, S. Stauffer, N. Katheder, P. Horvath, R. Mancini, A. Helenius, and M. Peter. 2012. "Cullin-3 regulates late endosome maturation." *Proceedings of the National Academy of Sciences of the United States of America* 109 (3): 823-8. https://doi.org/10.1073/pnas.1118744109.

Iaboni, M., V. Russo, R. Fontanella, G. Roscigno, D. Fiore, E. Donnarumma, C. L. Esposito, C. Quintavalle, P. H. Giangrande, V. de Franciscis, and G. Condorelli. 2016. "Aptamer-miRNA-212 Conjugate Sensitizes NSCLC Cells to TRAIL." *Mol Ther Nucleic Acids* 5: e289. https://doi.org/10.1038/mtna.2016.5.

Jauvin, D., J. Chretien, S. K. Pandey, L. Martineau, L. Revillod, G. Bassez, A. Lachon, A. R. MacLeod, G. Gourdon, T. M. Wheeler, C. A. Thornton, C. F. Bennett, and J. Puymirat. 2017. "Targeting DMPK with Antisense Oligonucleotide Improves Muscle Strength in Myotonic Dystrophy Type 1 Mice." *Mol Ther Nucleic Acids* 7: 465-474. https://doi.org/10.1016/j.omtn.2017.05.007.

Jorgensen, R., A. R. Merrill, and G. R. Andersen. 2006. "The life and death of translation elongation factor 2." *Biochem Soc Trans* 34 (Pt 1): 1-6. https://doi.org/10.1042/BST20060001.

Jovic, M., M. Sharma, J. Rahajeng, and S. Caplan. 2010. "The early endosome: a busy sorting station for proteins at the crossroads." *Histol Histopathol* 25 (1): 99-112. https://doi.org/10.14670/HH-25.99.

Juliano, R., M. R. Alam, V. Dixit, and H. Kang. 2008. "Mechanisms and strategies for effective delivery of antisense and siRNA oligonucleotides." *Nucleic Acids Res* 36 (12): 4158-71. https://doi.org/10.1093/nar/gkn342.

Kanwar, R., and M. E. Fortini. 2008. "The big brain aquaporin is required for endosome maturation and notch receptor trafficking." *Cell* 133 (5): 852-63. https://doi.org/10.1016/j.cell.2008.04.038.

Kim, E., Y. Jung, H. Choi, J. Yang, J. S. Suh, Y. M. Huh, K. Kim, and S. Haam. 2010. "Prostate cancer cell death produced by the co-delivery of Bcl-xL shRNA and doxorubicin using an aptamer-conjugated polyplex." *Biomaterials* 31 (16): 4592-9. https://doi.org/10. 1016/j.biomaterials.2010.02.030.

Kim, G. H., R. M. Dayam, A. Prashar, M. Terebiznik, and R. J. Botelho. 2014. "PIKfyve inhibition interferes with phagosome and endosome maturation in macrophages." *Traffic* 15 (10): 1143-63. https://doi.org/10.1111/tra.12199.

Kim, M., D. M. Kim, K. S. Kim, W. Jung, and D. E. Kim. 2018. "Applications of Cancer Cell-Specific Aptamers in Targeted Delivery of Anticancer Therapeutic Agents." *Molecules* 23 (4). https://doi.org/10.3390/molecules23040830.

Kim, Y. M., C. H. Jung, M. Seo, E. K. Kim, J. M. Park, S. S. Bae, and D. H. Kim. 2015. "mTORC1 phosphorylates UVRAG to negatively regulate autophagosome and endosome maturation." *Mol Cell* 57 (2): 207-18. https://doi.org/10.1016/j.molcel.2014.11.013.

Kotula, J. W., E. D. Pratico, X. Ming, O. Nakagawa, R. L. Juliano, and B. A. Sullenger. 2012. "Aptamer-mediated delivery of splice-switching oligonucleotides to the nuclei of cancer cells." *Nucleic Acid Ther* 22 (3): 187-95. https://doi.org/10.1089/nat.2012.0347.

Lai, W. Y., W. Y. Wang, Y. C. Chang, C. J. Chang, P. C. Yang, and K. Peck. 2014. "Synergistic inhibition of lung cancer cell invasion, tumor growth and angiogenesis using aptamer-siRNA chimeras." *Biomaterials* 35 (9): 2905-14. https://doi.org/10.1016/j.biomaterials.2013.12.054.

Law, F., J. H. Seo, Z. Wang, J. L. DeLeon, Y. Bolis, A. Brown, W. X. Zong, G. Du, and C. E. Rocheleau. 2017. "The VPS34 PI3K negatively regulates RAB-5 during endosome maturation." *Journal of cell science* 130 (12): 2007-2017. https://doi.org/10.1242/jcs.194746.

Li, C., W. Jiang, Q. Hu, L. C. Li, L. Dong, R. Chen, Y. Zhang, Y. Tang, J. B. Thrasher, C. B. Liu, and B. Li. 2016. "Enhancing DPYSL3 gene expression via a promoter-targeted small activating RNA approach suppresses cancer cell motility and metastasis." *Oncotarget* 7 (16): 22893-910. https://doi.org/10.18632/oncotarget.8290.

Li, L., J. Hou, X. Liu, Y. Guo, Y. Wu, L. Zhang, and Z. Yang. 2014. "Nucleolin-targeting liposomes guided by aptamer AS1411 for the delivery of siRNA for the treatment of malignant melanomas." *Biomaterials* 35 (12): 3840-50. https://doi.org/10.1016/j.biomaterials.2014.01.019.

Lin, Y., Q. Qiu, S. C. Gill, and S. D. Jayasena. 1994. "Modified RNA sequence pools for in vitro selection." *Nucleic Acids Res* 22 (24): 5229-34. https://doi.org/10.1093/nar/22.24.5229.

Lipi, F., S. Chen, M. Chakravarthy, S. Rakesh, and R. N. Veedu. 2016. "In vitro evolution of chemically-modified nucleic acid aptamers: Pros and cons, and comprehensive selection strategies." *RNA Biol* 13 (12): 1232-1245. https://doi.org/10.1080/15476286.2016.1236173.

Liu, H. Y., X. Yu, H. Liu, D. Wu, and J. X. She. 2016. "Co-targeting EGFR and survivin with a bivalent aptamer-dual siRNA chimera effectively suppresses prostate cancer." *Sci Rep* 6: 30346. https://doi.org/10.1038/srep30346.

Liu, N., C. Zhou, J. Zhao, and Y. Chen. 2012. "Reversal of paclitaxel resistance in epithelial ovarian carcinoma cells by a MUC1 aptamer-let-7i chimera." *Cancer Invest* 30 (8): 577-82. https://doi.org/10.3109/07357907.2012.707265.

Liu, R., M. Gong, X. Li, Y. Zhou, W. Gao, A. Tulpule, P. M. Chaudhary, J. Jung, and P. S. Gill. 2010. "Induction, regulation, and biologic function of Axl receptor tyrosine kinase in Kaposi sarcoma." *Blood* 116 (2): 297-305. https://doi.org/10.1182/blood-2009-12-257154.

Liu, T. T., T. S. Gomez, B. K. Sackey, D. D. Billadeau, and C. G. Burd. 2012. "Rab GTPase regulation of retromer-mediated cargo export during endosome maturation." *Mol Biol Cell* 23 (13): 2505-15. https://doi.org/10.1091/mbc.E11-11-0915.

McBride, J. L., R. L. Boudreau, S. Q. Harper, P. D. Staber, A. M. Monteys, I. Martins, B. L. Gilmore, H. Burstein, R. W. Peluso, B. Polisky, B. J. Carter, and B. L. Davidson. 2008. "Artificial miRNAs mitigate shRNA-mediated toxicity in the brain: implications for the therapeutic development of RNAi." *Proc Natl Acad Sci U S A* 105 (15): 5868-73. https://doi.org/10.1073/pnas.0801775105.

McNamara, J. O., 2nd, E. R. Andrechek, Y. Wang, K. D. Viles, R. E. Rempel, E. Gilboa, B. A. Sullenger, and P. H. Giangrande. 2006a. "Cell type-specific delivery of siRNAs with aptamer-siRNA chimeras." *Nat Biotechnol* 24 (8): 1005-15. https://doi.org/10.1038/nbt1223.

---. 2006b. "Cell type-specific delivery of siRNAs with aptamer-siRNA chimeras." *Nature Biotechnology* 24 (8): 1005-15. https://doi.org/10.1038/nbt1223.

Mi, J., Y. Liu, Z. N. Rabbani, Z. Yang, J. H. Urban, B. A. Sullenger, and B. M. Clary. 2010. "In vivo selection of tumor-targeting RNA motifs." *Nature chemical biology* 6 (1): 22-4. https://doi.org/10.1038/nchembio.277.

Mi, J., P. Ray, J. Liu, C. T. Kuan, J. Xu, D. Hsu, B. A. Sullenger, R. R. White, and B. M. Clary. 2016. "In Vivo Selection Against Human Colorectal Cancer Xenografts Identifies an Aptamer That Targets RNA Helicase Protein DHX9." *Molecular therapy. Nucleic acids* 5: e315. https://doi.org/10.1038/mtna.2016.27.

Moore, C. B., E. H. Guthrie, M. T. Huang, and D. J. Taxman. 2010. "Short hairpin RNA (shRNA): design, delivery, and assessment of gene knockdown." *Methods Mol Biol* 629: 141-58. https://doi.org/10.1007/978-1-60761-657-3_10.

Murphy, R. F. 1991. "Maturation models for endosome and lysosome biogenesis." *Trends Cell Biol* 1 (4): 77-82. http://www.ncbi.nlm.nih.gov/pubmed/14731781.

Neff, C. P., J. Zhou, L. Remling, J. Kuruvilla, J. Zhang, H. Li, D. D. Smith, P. Swiderski, J. J. Rossi, and R. Akkina. 2011a. "An aptamer-siRNA chimera suppresses HIV-1 viral loads and protects from helper CD4(+) T cell decline in humanized mice." *Science translational medicine* 3 (66): 66ra6. https://doi.org/10.1126/scitranslmed.3001581.

---. 2011b. "An aptamer-siRNA chimera suppresses HIV-1 viral loads and protects from helper CD4(+) T cell decline in humanized mice." *Sci Transl Med* 3 (66): 66ra6. https://doi.org/10.1126/scitranslmed.3001581.

Nguyen, Q., and T. Yokota. 2019. "Antisense oligonucleotides for the treatment of cardiomyopathy in Duchenne muscular dystrophy." *Am J Transl Res* 11 (3): 1202-1218. http://www.ncbi.nlm.nih.gov/pubmed/30972156.

Ni, S., H. Yao, L. Wang, J. Lu, F. Jiang, A. Lu, and G. Zhang. 2017. "Chemical Modifications of Nucleic Acid Aptamers for Therapeutic Purposes." *Int J Mol Sci* 18 (8). https://doi.org/10.3390/ijms18081683.

Ni, X., M. Castanares, A. Mukherjee, and S. E. Lupold. 2011. "Nucleic acid aptamers: clinical applications and promising new horizons." *Current medicinal chemistry* 18 (27): 4206-14. http://www.ncbi.nlm.nih.gov/pubmed/21838685.

Ni, X., Y. Zhang, J. Ribas, W. H. Chowdhury, M. Castanares, Z. Zhang, M. Laiho, T. L. DeWeese, and S. E. Lupold. 2011. "Prostate-targeted radiosensitization via aptamer-shRNA chimeras in human tumor xenografts." *J Clin Invest* 121 (6): 2383-90. https://doi.org/10.1172/JCI45109.

Ni, X., Y. Zhang, K. Zennami, M. Castanares, A. Mukherjee, R. R. Raval, H. Zhou, T. L. DeWeese, and S. E. Lupold. 2015. "Systemic Administration and Targeted Radiosensitization via Chemically Synthetic Aptamer-siRNA Chimeras in Human Tumor Xenografts." *Mol Cancer Ther* 14 (12): 2797-804. https://doi.org/10.1158/1535-7163.MCT-15-0291-T.

O'Sullivan, C. K. 2002. "Aptasensors--the future of biosensing?" *Analytical and bioanalytical chemistry* 372 (1): 44-8. https://doi.org/10.1007/s00216-001-1189-3.

Oliveto, S., M. Mancino, N. Manfrini, and S. Biffo. 2017. "Role of microRNAs in translation regulation and cancer." *World J Biol Chem* 8 (1): 45-56. https://doi.org/10.4331/wjbc.v8.i1.45.

Orava, E. W., N. Cicmil, and J. Gariepy. 2010. "Delivering cargoes into cancer cells using DNA aptamers targeting internalized surface portals." *Biochimica et biophysica acta* 1798 (12): 2190-200. https://doi.org/10.1016/j.bbamem.2010.02.004.

Otake, Y., S. Soundararajan, T. K. Sengupta, E. A. Kio, J. C. Smith, M. Pineda-Roman, R. K. Stuart, E. K. Spicer, and D. J. Fernandes. 2007. "Overexpression of nucleolin in chronic lymphocytic leukemia cells induces stabilization of bcl2 mRNA." *Blood* 109 (7): 3069-75. https://doi.org/10.1182/blood-2006-08-043257.

Pang, K. M., D. Castanotto, H. Li, L. Scherer, and J. J. Rossi. 2018. "Incorporation of aptamers in the terminal loop of shRNAs yields an effective and novel combinatorial targeting strategy." *Nucleic Acids Res* 46 (1): e6. https://doi.org/10.1093/nar/gkx980.

Pastor, F., D. Kolonias, J. O. McNamara, 2nd, and E. Gilboa. 2011. "Targeting 4-1BB costimulation to disseminated tumor lesions with bi-specific oligonucleotide aptamers." *Mol Ther* 19 (10): 1878-86. https://doi.org/10.1038/mt.2011.145.

Philippou, S., N. P. Mastroyiannopoulos, N. Makrides, C. W. Lederer, M. Kleanthous, and L. A. Phylactou. 2018. "Selection and Identification of Skeletal-Muscle-Targeted RNA Aptamers." *Mol Ther Nucleic Acids* 10: 199-214. https://doi.org/10.1016/j.omtn.2017.12.004.

Pofahl, M., J. Wengel, and G. Mayer. 2014. "Multifunctional nucleic acids for tumor cell treatment." *Nucleic Acid Ther* 24 (2): 171-7. https://doi.org/10.1089/nat.2013.0472.

Puchner, E. M., J. M. Walter, R. Kasper, B. Huang, and W. A. Lim. 2013. "Counting molecules in single organelles with superresolution microscopy allows tracking of the endosome maturation trajectory." *Proceedings of the National Academy of Sciences of the United States of America* 110 (40): 16015-20. https://doi.org/10.1073/pnas.1309676110.

Qiu, W., F. Zhou, Q. Zhang, X. Sun, X. Shi, Y. Liang, X. Wang, and L. Yue. 2013. "Overexpression of nucleolin and different expression sites both related to the prognosis of gastric cancer." *APMIS* 121 (10): 919-25. https://doi.org/10.1111/apm.12131.

Ranches, G., M. Lukasser, H. Schramek, A. Ploner, T. Stasyk, G. Mayer, G. Mayer, and A. Huttenhofer. 2017. "In Vitro Selection of Cell-Internalizing DNA Aptamers in a Model System of Inflammatory Kidney Disease." *Mol Ther Nucleic Acids* 8: 198-210. https://doi.org/10.1016/j.omtn.2017.06.018.

Relizani, K., G. Griffith, L. Echevarria, F. Zarrouki, P. Facchinetti, C. Vaillend, C. Leumann, L. Garcia, and A. Goyenvalle. 2017. "Efficacy and Safety Profile of Tricyclo-DNA Antisense Oligonucleotides in Duchenne Muscular Dystrophy Mouse Model." *Mol Ther Nucleic Acids* 8: 144-157. https://doi.org/10.1016/j.omtn.2017.06.013.

Rockey, W. M., F. J. Hernandez, S. Y. Huang, S. Cao, C. A. Howell, G. S. Thomas, X. Y. Liu, N. Lapteva, D. M. Spencer, J. O. McNamara, X. Zou, S. J. Chen, and P. H. Giangrande. 2011. "Rational truncation of an RNA aptamer to prostate-specific membrane antigen using computational structural modeling." *Nucleic Acid Ther* 21 (5): 299-314. https://doi.org/10.1089/nat.2011.0313.

Rohde, J. H., J. E. Weigand, B. Suess, and S. Dimmeler. 2015. "A Universal Aptamer Chimera for the Delivery of Functional microRNA-126." *Nucleic Acid Ther* 25 (3): 141-51. https://doi.org/10.1089/nat.2014.0501.

Russo, V., A. Paciocco, A. Affinito, G. Roscigno, D. Fiore, F. Palma, M. Galasso, S. Volinia, A. Fiorelli, C. L. Esposito, S. Nuzzo, G. Inghirami, V. de Franciscis, and G. Condorelli. 2018. "Aptamer-miR-34c Conjugate Affects Cell Proliferation of Non-Small-Cell Lung Cancer Cells." *Mol Ther Nucleic Acids* 13: 334-346. https://doi.org/10.1016/j.omtn.2018.09.016.

Sakai, Y., M. S. Islam, M. Adamiak, S. C. Shiu, J. A. Tanner, and J. G. Heddle. 2018. "DNA Aptamers for the Functionalisation of DNA

Origami Nanostructures." *Genes (Basel)* 9 (12). https://doi.org/10.3390/genes9120571.

Sanchez-Luque, F. J., M. Stich, S. Manrubia, C. Briones, and A. Berzal-Herranz. 2014. "Efficient HIV-1 inhibition by a 16 nt-long RNA aptamer designed by combining in vitro selection and in silico optimisation strategies." *Sci Rep* 4: 6242. https://doi.org/10.1038/srep06242.

Schrand, B., A. Berezhnoy, R. Brenneman, A. Williams, A. Levay, and E. Gilboa. 2015. "Reducing toxicity of 4-1BB costimulation: targeting 4-1BB ligands to the tumor stroma with bi-specific aptamer conjugates." *Oncoimmunology* 4 (3): e970918. https://doi.org/10.4161/21624011.2014.970918.

Schrand, B., A. Berezhnoy, R. Brenneman, A. Williams, A. Levay, L. Y. Kong, G. Rao, S. Zhou, A. B. Heimberger, and E. Gilboa. 2014. "Targeting 4-1BB costimulation to the tumor stroma with bispecific aptamer conjugates enhances the therapeutic index of tumor immunotherapy." *Cancer Immunol Res* 2 (9): 867-77. https://doi.org/10.1158/2326-6066.CIR-14-0007.

Scott, C. C., F. Vacca, and J. Gruenberg. 2014. "Endosome maturation, transport and functions." *Semin Cell Dev Biol* 31: 2-10. https://doi.org/10.1016/j.semcdb.2014.03.034.

Serfass, J. M., Y. Takahashi, Z. Zhou, Y. I. Kawasawa, Y. Liu, N. Tsotakos, M. M. Young, Z. Tang, L. Yang, J. M. Atkinson, Z. C. Chroneos, and H. G. Wang. 2017. "Endophilin B2 facilitates endosome maturation in response to growth factor stimulation, autophagy induction, and influenza A virus infection." *The Journal of biological chemistry* 292 (24): 10097-10111. https://doi.org/10.1074/jbc.M117.792747.

Shah, A. H., N. L. Cianciola, J. L. Mills, F. D. Sonnichsen, and C. Carlin. 2007. "Adenovirus RIDalpha regulates endosome maturation by mimicking GTP-Rab7." *J Cell Biol* 179 (5): 965-80. https://doi.org/10.1083/jcb.200702187.

Shangguan, D., L. Meng, Z. C. Cao, Z. Xiao, X. Fang, Y. Li, D. Cardona, R. P. Witek, C. Liu, and W. Tan. 2008. "Identification of liver cancer-specific aptamers using whole live cells." *Analytical chemistry* 80 (3): 721-8. https://doi.org/10.1021/ac701962v.

Shen, X., and D. R. Corey. 2018. "Chemistry, mechanism and clinical status of antisense oligonucleotides and duplex RNAs." *Nucleic Acids Res* 46 (4): 1584-1600. https://doi.org/10.1093/nar/gkx1239.

Shih, J. Y., M. F. Tsai, T. H. Chang, Y. L. Chang, A. Yuan, C. J. Yu, S. B. Lin, G. Y. Liou, M. L. Lee, J. J. Chen, T. M. Hong, S. C. Yang, J. L. Su, Y. C. Lee, and P. C. Yang. 2005. "Transcription repressor slug promotes carcinoma invasion and predicts outcome of patients with lung adenocarcinoma." *Clin Cancer Res* 11 (22): 8070-8. https://doi.org/10.1158/1078-0432.CCR-05-0687.

Sivakumar, P., S. Kim, H. C. Kang, and M. S. Shim. 2019. "Targeted siRNA delivery using aptamer-siRNA chimeras and aptamer-conjugated nanoparticles." *Wiley Interdiscip Rev Nanomed Nanobiotechnol* 11 (3): e1543. https://doi.org/10.1002/wnan.1543.

Soldevilla, M. M., D. Meraviglia-Crivelli de Caso, A. P. Menon, and F. Pastor. 2018. "Aptamer-iRNAs as Therapeutics for Cancer Treatment." *Pharmaceuticals (Basel)* 11 (4). https://doi.org/10.3390/ph11040108.

Solinger, J. A., and A. Spang. 2014. "Loss of the Sec1/Munc18-family proteins VPS-33.2 and VPS-33.1 bypasses a block in endosome maturation in Caenorhabditis elegans." *Mol Biol Cell* 25 (24): 3909-25. https://doi.org/10.1091/mbc.E13-12-0710.

Stenvang, J., A. Petri, M. Lindow, S. Obad, and S. Kauppinen. 2012. "Inhibition of microRNA function by antimiR oligonucleotides." *Silence* 3 (1): 1. https://doi.org/10.1186/1758-907X-3-1.

Sun, H., X. Zhu, P. Y. Lu, R. R. Rosato, W. Tan, and Y. Zu. 2014. "Oligonucleotide aptamers: new tools for targeted cancer therapy." *Molecular therapy. Nucleic acids* 3: e182. https://doi.org/10.1038/mtna.2014.32.

Szeto, K., D. R. Latulippe, A. Ozer, J. M. Pagano, B. S. White, D. Shalloway, J. T. Lis, and H. G. Craighead. 2013. "RAPID-SELEX for RNA aptamers." *PloS one* 8 (12): e82667. https://doi.org/10.1371/journal.pone.0082667.

Takahashi, Y., S. Nada, S. Mori, T. Soma-Nagae, C. Oneyama, and M. Okada. 2012. "The late endosome/lysosome-anchored p18-mTORC1 pathway controls terminal maturation of lysosomes." *Biochemical and biophysical research communications* 417 (4): 1151-7. https://doi.org/10.1016/j.bbrc.2011.12.082.

Tanno, T., P. Zhang, C. A. Lazarski, Y. Liu, and P. Zheng. 2017. "An aptamer-based targeted delivery of miR-26a protects mice against chemotherapy toxicity while suppressing tumor growth." *Blood Adv* 1 (15): 1107-1119. https://doi.org/10.1182/bloodadvances.2017004705.

Tawiah, K. D., D. Porciani, and D. H. Burke. 2017. "Toward the Selection of Cell Targeting Aptamers with Extended Biological Functionalities to Facilitate Endosomal Escape of Cargoes." *Biomedicines* 5 (3). https://doi.org/10.3390/biomedicines5030051.

Thiel, K. W., L. I. Hernandez, J. P. Dassie, W. H. Thiel, X. Liu, K. R. Stockdale, A. M. Rothman, F. J. Hernandez, J. O. McNamara, 2nd, and P. H. Giangrande. 2012a. "Delivery of chemo-sensitizing siRNAs to HER2+-breast cancer cells using RNA aptamers." *Nucleic acids research* 40 (13): 6319-37. https://doi.org/10.1093/nar/gks294.

---. 2012b. "Delivery of chemo-sensitizing siRNAs to HER2+-breast cancer cells using RNA aptamers." *Nucleic Acids Res* 40 (13): 6319-37. https://doi.org/10.1093/nar/gks294.

Thiel, W. H., T. Bair, A. S. Peek, X. Liu, J. Dassie, K. R. Stockdale, M. A. Behlke, F. J. Miller, Jr., and P. H. Giangrande. 2012. "Rapid identification of cell-specific, internalizing RNA aptamers with bioinformatics analyses of a cell-based aptamer selection." *PLoS One* 7 (9): e43836. https://doi.org/10.1371/journal.pone.0043836.

Tuerk, C., and L. Gold. 1990. "Systematic evolution of ligands by exponential enrichment: RNA ligands to bacteriophage T4 DNA polymerase." *Science* 249 (4968): 505-10.

van Putten, M., C. Tanganyika-de Winter, S. Bosgra, and A. Aartsma-Rus. 2019. "Nonclinical Exon Skipping Studies with 2'-O-Methyl Phosphorothioate Antisense Oligonucleotides in mdx and mdx-utrn-/- Mice Inspired by Clinical Trial Results." *Nucleic Acid Ther* 29 (2): 92-103. https://doi.org/10.1089/nat.2018.0759.

van Weering, J. R., P. Verkade, and P. J. Cullen. 2012. "SNX-BAR-mediated endosome tubulation is co-ordinated with endosome maturation." *Traffic* 13 (1): 94-107. https://doi.org/10.1111/j.1600-0854.2011.01297.x.

Varkouhi, A. K., M. Scholte, G. Storm, and H. J. Haisma. 2011. "Endosomal escape pathways for delivery of biologicals." *Journal of controlled release : official journal of the Controlled Release Society* 151 (3): 220-8. https://doi.org/10.1016/j.jconrel.2010.11.004.

Wallner, C., M. Drysch, M. Becerikli, H. Jaurich, J. M. Wagner, S. Dittfeld, J. Nagler, K. Harati, M. Dadras, S. Philippou, M. Lehnhardt, and B. Behr. 2017. "Interaction with the GDF8/11 pathway reveals treatment options for adenocarcinoma of the breast." *Breast* 37: 134-141. https://doi.org/10.1016/j.breast.2017.11.010.

Wang, H., Y. Zhang, H. Yang, M. Qin, X. Ding, R. Liu, and Y. Jiang. 2018. "In Vivo SELEX of an Inhibitory NSCLC-Specific RNA Aptamer from PEGylated RNA Library." *Mol Ther Nucleic Acids* 10: 187-198. https://doi.org/10.1016/j.omtn.2017.12.003.

Wang, Y., X. Chen, B. Tian, J. Liu, L. Yang, L. Zeng, T. Chen, A. Hong, and X. Wang. 2017. "Nucleolin-targeted Extracellular Vesicles as a Versatile Platform for Biologics Delivery to Breast Cancer." *Theranostics* 7 (5): 1360-1372. https://doi.org/10.7150/thno.16532.

Wang, Y., Y. Luo, T. Bing, Z. Chen, M. Lu, N. Zhang, D. Shangguan, and X. Gao. 2014. "DNA aptamer evolved by cell-SELEX for recognition of prostate cancer." *PloS one* 9 (6): e100243. https://doi.org/10.1371/journal.pone.0100243.

Wang, Y., B. Yao, Y. Wang, M. Zhang, S. Fu, H. Gao, R. Peng, L. Zhang, and J. Tang. 2014. "Increased FoxM1 expression is a target for metformin in the suppression of EMT in prostate cancer." *Int J Mol Med* 33 (6): 1514-22. https://doi.org/10.3892/ijmm.2014.1707.

Wavre-Shapton, S. T., I. P. Meschede, M. C. Seabra, and C. E. Futter. 2014. "Phagosome maturation during endosome interaction revealed by partial rhodopsin processing in retinal pigment epithelium." *Journal of cell science* 127 (Pt 17): 3852-61. https://doi.org/10.1242/jcs.154757.

Wen, H., H. Jung, and X. Li. 2015. "Drug Delivery Approaches in Addressing Clinical Pharmacology-Related Issues: Opportunities and Challenges." *AAPS J* 17 (6): 1327-40. https://doi.org/10.1208/s12248-015-9814-9.

Wheeler, L. A., R. Trifonova, V. Vrbanac, E. Basar, S. McKernan, Z. Xu, E. Seung, M. Deruaz, T. Dudek, J. I. Einarsson, L. Yang, T. M. Allen, A. D. Luster, A. M. Tager, D. M. Dykxhoorn, and J. Lieberman. 2011. "Inhibition of HIV transmission in human cervicovaginal explants and humanized mice using CD4 aptamer-siRNA chimeras." *J Clin Invest* 121 (6): 2401-12. https://doi.org/10.1172/JCI45876.

White, R. R., B. A. Sullenger, and C. P. Rusconi. 2000. "Developing aptamers into therapeutics." *The Journal of clinical investigation* 106 (8): 929-34. https://doi.org/10.1172/JCI11325.

Wilson, J. M., M. de Hoop, N. Zorzi, B. H. Toh, C. G. Dotti, and R. G. Parton. 2000. "EEA1, a tethering protein of the early sorting endosome, shows a polarized distribution in hippocampal neurons, epithelial cells, and fibroblasts." *Mol Biol Cell* 11 (8): 2657-71. http://www.ncbi.nlm.nih.gov/pubmed/10930461.

Wu, M., H. Zhao, L. Guo, Y. Wang, J. Song, X. Zhao, C. Li, L. Hao, D. Wang, and J. Tang. 2018. "Ultrasound-mediated nanobubble destruction (UMND) facilitates the delivery of A10-3.2 aptamer targeted and siRNA-loaded cationic nanobubbles for therapy of prostate cancer." *Drug Deliv* 25 (1): 226-240. https://doi.org/10.1080/10717544.2017.1422300.

Wu, X., B. Ding, J. Gao, H. Wang, W. Fan, X. Wang, W. Zhang, X. Wang, L. Ye, M. Zhang, X. Ding, J. Liu, Q. Zhu, and S. Gao. 2011. "Second-generation aptamer-conjugated PSMA-targeted delivery system for prostate cancer therapy." *Int J Nanomedicine* 6: 1747-56. https://doi.org/10.2147/IJN.S23747.

Wullner, U., I. Neef, A. Eller, M. Kleines, M. K. Tur, and S. Barth. 2008. "Cell-specific induction of apoptosis by rationally designed bivalent aptamer-siRNA transcripts silencing eukaryotic elongation factor 2." *Curr Cancer Drug Targets* 8 (7): 554-65. http://www.ncbi.nlm.nih.gov/pubmed/18991566.

Wurster, C. D., and A. C. Ludolph. 2018. "Antisense oligonucleotides in neurological disorders." *Ther Adv Neurol Disord* 11: 1756286418776932. https://doi.org/10.1177/1756286418776932.

Xu, X., J. Wu, Y. Liu, P. E. Saw, W. Tao, M. Yu, H. Zope, M. Si, A. Victorious, J. Rasmussen, D. Ayyash, O. C. Farokhzad, and J. Shi. 2017. "Multifunctional Envelope-Type siRNA Delivery Nanoparticle Platform for Prostate Cancer Therapy." *ACS Nano* 11 (3): 2618-2627. https://doi.org/10.1021/acsnano.6b07195.

Yan, A. C., and M. Levy. 2009. "Aptamers and aptamer targeted delivery." *RNA biology* 6 (3): 316-20.

Yan, A., and M. Levy. 2014. "Cell internalization SELEX: in vitro selection for molecules that internalize into cells." *Methods in molecular biology* 1103: 241-65. https://doi.org/10.1007/978-1-62703-730-3_18.

Yang, S., Z. Ren, M. Chen, Y. Wang, B. You, W. Chen, C. Qu, Y. Liu, and X. Zhang. 2018. "Nucleolin-Targeting AS1411-Aptamer-Modified Graft Polymeric Micelle with Dual pH/Redox Sensitivity Designed To Enhance Tumor Therapy through the Codelivery of Doxorubicin/TLR4 siRNA and Suppression of Invasion." *Mol Pharm* 15 (1): 314-325. https://doi.org/10.1021/acs.molpharmaceut.7b01093.

Yoon, S., K. W. Huang, V. Reebye, P. Mintz, Y. W. Tien, H. S. Lai, P. Saetrom, I. Reccia, P. Swiderski, B. Armstrong, A. Jozwiak, D. Spalding, L. Jiao, N. Habib, and J. J. Rossi. 2016. "Targeted Delivery of C/EBPalpha -saRNA by Pancreatic Ductal Adenocarcinoma-specific RNA Aptamers Inhibits Tumor Growth In Vivo." *Mol Ther* 24 (6): 1106-1116. https://doi.org/10.1038/mt.2016.60.

Yoon, S., and J. J. Rossi. 2018. "Aptamers: Uptake mechanisms and intracellular applications." *Adv Drug Deliv Rev* 134: 22-35. https://doi.org/10.1016/j.addr.2018.07.003.

Zhang, Z., M. Blank, and H. J. Schluesener. 2004. "Nucleic acid aptamers in human viral disease." *Arch Immunol Ther Exp (Warsz)* 52 (5): 307-15. http://www.ncbi.nlm.nih.gov/pubmed/15507871.

Zhou, J., M. L. Bobbin, J. C. Burnett, and J. J. Rossi. 2012. "Current progress of RNA aptamer-based therapeutics." *Front Genet* 3: 234. https://doi.org/10.3389/fgene.2012.00234.

Zhou, J., D. Lazar, H. Li, X. Xia, S. Satheesan, P. Charlins, D. O'Mealy, R. Akkina, S. Saayman, M. S. Weinberg, J. J. Rossi, and K. V. Morris. 2018. "Receptor-targeted aptamer-siRNA conjugate-directed transcriptional regulation of HIV-1." *Theranostics* 8 (6): 1575-1590. https://doi.org/10.7150/thno.23085.

Zhou, J., H. Li, S. Li, J. Zaia, and J. J. Rossi. 2008a. "Novel dual inhibitory function aptamer-siRNA delivery system for HIV-1 therapy." *Molecular therapy : the journal of the American Society of Gene Therapy* 16 (8): 1481-9. https://doi.org/10.1038/mt.2008.92.

---. 2008b. "Novel dual inhibitory function aptamer-siRNA delivery system for HIV-1 therapy." *Mol Ther* 16 (8): 1481-9. https://doi.org/10.1038/mt.2008.92.

Zhou, J., C. P. Neff, P. Swiderski, H. Li, D. D. Smith, T. Aboellail, L. Remling-Mulder, R. Akkina, and J. J. Rossi. 2013. "Functional in vivo delivery of multiplexed anti-HIV-1 siRNAs via a chemically synthesized aptamer with a sticky bridge." *Mol Ther* 21 (1): 192-200. https://doi.org/10.1038/mt.2012.226.

Zhou, J., and J. Rossi. 2016. "Aptamers as targeted therapeutics: current potential and challenges." *Nature reviews. Drug discovery*. https://doi.org/10.1038/nrd.2016.199.

---. 2017. "Aptamers as targeted therapeutics: current potential and challenges." *Nature reviews. Drug discovery* 16 (6): 440. https://doi.org/10.1038/nrd.2017.86.

Zhou, J., and J. J. Rossi. 2011. "Cell-specific aptamer-mediated targeted drug delivery." *Oligonucleotides* 21 (1): 1-10. https://doi.org/10.1089/oli.2010.0264.

---. 2014a. "Cell-type-specific, Aptamer-functionalized Agents for Targeted Disease Therapy." *Molecular therapy. Nucleic acids* 3: e169. https://doi.org/10.1038/mtna.2014.21.

Zhou, J., J. J. Rossi, and K. T. Shum. 2015. "Methods for assembling B-cell lymphoma specific and internalizing aptamer-siRNA nanoparticles via the sticky bridge." *Methods Mol Biol* 1297: 169-85. https://doi.org/10.1007/978-1-4939-2562-9_12.

Zhou, J., S. Satheesan, H. Li, M. S. Weinberg, K. V. Morris, J. C. Burnett, and J. J. Rossi. 2015. "Cell-specific RNA aptamer against human CCR5 specifically targets HIV-1 susceptible cells and inhibits HIV-1 infectivity." *Chem Biol* 22 (3): 379-90. https://doi.org/10.1016/j.chembiol.2015.01.005.

Zhou, J., P. Swiderski, H. Li, J. Zhang, C. P. Neff, R. Akkina, and J. J. Rossi. 2009a. "Selection, characterization and application of new RNA HIV gp 120 aptamers for facile delivery of Dicer substrate siRNAs into HIV infected cells." *Nucleic Acids Res* 37 (9): 3094-109. https://doi.org/10.1093/nar/gkp185.

---. 2009b. "Selection, characterization and application of new RNA HIV gp 120 aptamers for facile delivery of Dicer substrate siRNAs into HIV infected cells." *Nucleic acids research* 37 (9): 3094-109. https://doi.org/10.1093/nar/gkp185.

Zhou, Jiehua, and John Rossi. 2014b. "Cell-type-specific aptamer and aptamer-small interfering RNA conjugates for targeted human immunodeficiency virus type 1 therapy." *Journal of investigative medicine : the official publication of the American Federation for Clinical Research* 62 (7): 914-919. https://doi.org/10.1097/jim.0000000000000103.

Zhu, Q., T. Shibata, T. Kabashima, and M. Kai. 2012. "Inhibition of HIV-1 protease expression in T cells owing to DNA aptamer-mediated specific delivery of siRNA." *Eur J Med Chem* 56: 396-9. https://doi.org/10.1016/j.ejmech.2012.07.045.

Biographical Sketch

Leonidas A. Phylactou

Affiliation: Department of Molecular Genetics, Function & Therapy (MGFT), The Cyprus Institute of Neurology and Genetics (CING), Nicosia, Cyprus.

Education: PhD in Molecular Genetics

Business Address: The Cyprus Institute of Neurology & Genetics, PO Box 23462, 1683 Nicosia, Cyprus.

Research and Professional Experience: My research experience has been mainly focused on developing therapies, biomarkers and studying mechanisms in muscular dystrophy. These include regulatory RNAs (miRNAs, siRNA, antisense oligonucleotides). I have received funding for the above activities from several international and local funding bodies such as the Association Francaise Contres Les Myopathies, Muscular Dystrophy Campaign UK, the A.G. Leventis Foundation and the Cyprus Research Promotion Foundation. I have published extensively in the above areas in high impact factor journals such as Nature Genetics, EMBO Reports, Human Molecular Genetics and Journal of Cell Science. Other publications in which I have been involved either as a senior author or as a co-author are related mostly to inherited diseases. I have several international collaborations on the fields of regulatory RNA and muscular dystrophy and participated in the European Union network "Networking towards clinical application of antisense-mediated exon skipping" COST BM1207 and currently in COST CA17103 "Delivery of Antisense RNA Therapeutics." Furthermore, I have shown that I can manage a laboratory and staff under research projects but also diagnostic services to patients. Being at the Cyprus Institute of Neurology and Genetics (CING), an organization in which clinical services co-exist with research on translational medicine, provides an advantage since patients, patient groups, diagnostic laboratories and pre-clinical and clinical research are all under the same roof. Being also the Chief Executive of CING allows me to have a global view on the fields of neurology and genetics which are directly related to the project and a very good understanding of the problems patients are facing.

Professional Appointments:

1/2/2015 – Senior Scientist, Head of Department of Molecular Genetics, Function and Therapy, CING.

1/2/2005 – Member of the Scientific Council, CING.

02/2012 – Professor, Cyprus School of Molecular Medicine, CING.

11/2015 – Chairman of the Scientific Council, CING.

11/2015 – Provost, Cyprus School of Molecular Medicine, CING.

11/2015 – Chief Executive Medical Director, CING.

Honors:

1997-98 – Elected Fulford Junior Research Fellow, Somerville College, Oxford.

2011-13 – Elected President of the Cyprus Society of Human Genetics.

2014 – Leader of the Greek Cypriot Bicommunal Technical Health Committee-appointed by the President of The Republic of Cyprus.

Publications from the Last 3 Years:

1) A novel heterozygous duplication of the SLC12A3 gene in two Gitelman syndrome pedigrees: indicating a founder effect. Fanis P, Efstathiou E, Neocleous V, **Phylactou LA**, Hadjipanayis A. *J Genet.* 2019 Mar,98(1). pii.22.
2) Genotype Is Associated to the Degree of Virilization in Patients With Classic Congenital Adrenal Hyperplasia. Neocleous V, Fanis P, **Phylactou LA**, Skordis N. Front Endocrinol (Lausanne). 2018 Dec 3;9:733. doi: 10.3389/fendo.2018.00733. eCollection 2018. PMID: 30559721.
3) Multiple Endocrine Neoplasia 2 in Cyprus: Evidence for a founder effect. P. Fanis, N. Skordis, S. Frangos, G. Christopoulos, E. Spanou-Aristidou, E. Andreou, P. Manoli, M. Mavrommatis, S. Nicolaou, M. Kleanthous, M. A. Cariolou, V. Christophidou-Anastasiadou, G. A. Tanteles, **L. A. Phylactou**, V. Neocleous. *J Endocrinol Invest* (2018). https://doi.org/10.1007/s40618-018-0841-0.

4) Selection and Identification of Skeletal-Muscle-Targeted RNA Aptamers. Philippou S, Mastroyiannopoulos NP, Makrides N, Lederer CW, Kleanthous M, **Phylactou LA**. *Mol Ther Nucleic Acids.* 2018 Mar 2;10:199-214. doi: 10.1016/j.omtn.2017.12.004. Epub 2017 Dec 9. PMID: 29499933.

5) GnRH-dependent precocious puberty manifested at the age of 14 months in a girl with 47,XXX karyotype. Skordis N, Ferrari E, Antoniadou A, **Phylactou LA**, Fanis P, Neocleous V. *Hormones* (Athens). 2017 Jul;16(3):318-321. doi: 10.14310/horm.2002.1740. PMID: 29278519.

6) Identification of exosomal muscle-specific miRNAs in serum of myotonic dystrophy patients relating to muscle disease progress. Koutsoulidou A, Photiades M, Kyriakides TC, Georgiou K, Prokopi M, Kapnisis K, Lusakowska A, Nearchou M, Christou Y, Papadimas GK, Anayiotos A, Kyriakou K, Kararizou E, Zamba Papanicolaou E, **Phylactou LA.** *Hum Mol Genet.* 2017 Sep 1;26(17):3285-3302. doi: 10.1093/hmg/ddx212. PMID: 28637233.

7) Variations in the 3'UTR of the *CYP21A2* Gene in Heterozygous Females with Hyperandrogenaemia. Neocleous V, Fanis P, Toumba M, Phedonos AAP, Picolos M, Andreou E, Kyriakides TC, Tanteles GA, Shammas C, **Phylactou LA**, Skordis N. *Int J Endocrinol.* 2017;2017:8984365. doi: 10.1155/2017/8984365. Epub 2017 Apr 12. PMID: 28487735.

8) Successful use of tocilizumab in two cases of severe autoinflammatory disease with a single copy of the Mediterranean fever gene. Nikiphorou E, Neocleous V, **Phylactou LA,** Psarelis S. *Rheumatology* (Oxford). 2017 Sep 1;56(9):1627-1628. doi: 10.1093/rheumatology/kex180. PMID: 28486679.

9) CDKN2A and MC1R variants found in Cypriot patients diagnosed with cutaneous melanoma. Koulermou G, Shammas C, Vassiliou A, Kyriakides TC, Costi C, Neocleous V, **Phylactou LA**, Pantelidou M. *J Genet.* 2017 Mar;96(1):155-160. PMID: 28360400.

10) A novel MKRN3 nonsense mutation causing familial central precocious puberty. Christoforidis A, Skordis N, Fanis P, Dimitriadou

M, Sevastidou M, Phelan MM, Neocleous V, **Phylactou LA.** *Endocrine.* 2017 May;56(2):446-449. doi: 10.1007/s12020-017-1232-6. Epub 2017 Jan 28. PMID: 28132164.

11) Identification of a novel 15.5 kb SHOX deletion associated with marked intrafamilial phenotypic variability and analysis of its molecular origin. Alexandrou A, Papaevripidou I, Tsangaras K, Alexandrou I, Tryfonidis M, ChristophidouAnastasiadou V, Zamba-Papanicolaou E, Koumbaris G, Neocleous V, **Phylactou LA**, Skordis N, Tanteles GA, Sismani C. *J Genet.* 2016 Dec;95(4):839-845. PMID: 27994182.
12) Evidence of digenic inheritance in autoinflammation-associated genes. Neocleous V, Byrou S, Toumba M, Costi C, Shammas C, Kyriakou C, ChristophidouAnastasiadou V, Tanteles GA, Hadjipanayis A, **Phylactou LA**. *J Genet.* 2016 Dec;95(4):761-766. PMID: 27994174.
13) A novel MC4R deletion coexisting with FTO and MC1R gene variants, causes severe early onset obesity. Neocleous V, Shammas C, Phelan MM, Fanis P, Pantelidou M, Skordis N, Mantzoros C, **Phylactou LA**, Toumba M. *Hormones* (Athens). 2016 Jul;15(3):445-452. doi: 10.14310/horm.2002.1686. PMID: 27394708.
14) Genetic screening of non-classic CAH females with hyperandrogenemia identifies a novel CYP11B1 gene mutation. Shammas C, Byrou S, Phelan MM, Toumba M, Stylianou C, Skordis N, Neocleous V**, Phylactou LA**. *Hormones* (Athens). 2016 Apr;15(2):235-242. doi: 10.14310/horm.2002.1651. PMID: 27376426.

In: A Comprehensive Guide to Aptamers
Editor: Tom Shuster
ISBN: 978-1-53616-293-6

Chapter 2

APTAMERS AS RADIOPHARMACEUTICALS

Renata Salgado Fernandes[1], André Luís Branco de Barros[1] and Antero Silva Ribeiro de Andrade[2,*]

[1]Departamento de Análises Clínicas e Toxicológicas, Faculdade de Farmácia, Universidade Federal de Minas Gerais (UFMG) Belo Horizonte, MG – Brazil

[2]Unidade de Radiobiologia, Centro de Desenvolvimento da Tecnologia Nuclear (CDTN), Comissão Nacional de Energia Nuclear (CNEN), Belo Horizonte, MG – Brasil

ABSTRACT

Acid nucleic aptamers are oligonucleotides that bind to a specific target molecule with high affinity and specificity. Because of their unique characteristics, aptamers are promising tools for development of new radiopharmaceuticals. They seem to be non-toxic and non-immunogenic, have small size and fast clearance. Aptamers can be selected for almost any target, including toxic and non-imunogenic molecules. Due to their small size and structural flexibility, aptamers may bind hidden epitopes. They

* Corresponding Author's E-mail: antero@cdtn.br.

can be easily produced by *in vitro* conditions with high reproducibility and free of contaminants, and the chemical synthesis makes them receptive to many modifications such as to make them more resistant to nucleases or to incorporate chelating groups. Aptamers can be labeled with different radioisotopes, thus allowing its application for imaging, therapy or theranostics, according to the radionuclide used. This chapter deals with the *in vivo* studies that were conducted with radiopharmaceuticals based on aptamers, the radionuclides used, the radiolabeling strategies, the chemical modifications of interest to improve their properties, and the main aptamers advantages and drawbacks for application as radiopharmaceuticals.

Keywords: aptamers, radiopharmaceuticals, nuclear medicine, radionuclide

1. Introduction

1.1. Radiopharmaceuticals

Nuclear medicine has begun in the 1950s, with sodium iodine ($Na^{131}I$) as the first radiopharmaceutical approved, by the FDA, for clinical use in thyroid diseases. Nowadays, the radioisotope ^{99m}Tc is the most used in diagnostic procedures, for a variety of diseases (GIJS, 2016). A radiopharmaceutical, typically, is based on two components: a radioactive element – the component that provides a signal that can be detected by an imaging instrument - and a pharmaceutical - a ligand that can specifically interact with the molecular target of interest. Although, in some cases, the radioactive element itself can be considered a radiopharmaceutical, e.g., ^{131}I, $^{99m}TcO_4^-$. Radiopharmaceuticals usually have a minimal pharmacologic effect, because in most cases, the pharmaceutical element is used in trace quantities. However, the usefulness of a radiopharmaceutical is dictated by its preferential localization in a given organ or its participation in the physiologic function of the organ (Saha, 2010; Muller-Bouvier, 2018).

An ideal radiopharmaceutical has to present some preferential characteristics such as: 1) be easily produced – complicated methods of production of radionuclides or labeled compounds increase the cost of the radiopharmaceutical; 2) short effective half-life – the radioisotope in use has to have a half-life long enough to complete the study in question, but not too long that exposes the patient to an unnecessary dose; 3) gamma(γ)-ray or positron emission – radionuclides decaying by alpha (α) or beta (β^-) are not used for diagnostic purposes, these radiation cause tissue damage, being used, thus, for treatment purposes. Moreover, is preferably that the radionuclide emits a γ radiation with an energy between 100 and 300 keV for SPECT imaging; 4) high target-to-non-target ratio – in the body, is desirable that the labeled compound accumulates preferentially in the organ under study since the activity from non-target areas can obscure the structural details of the images of the target organ; 5) renal elimination route – renal route is preferable over hepatobiliary clearance because it is accompanied by a gastrointestinal transit, making the process slower. To achieve renal clearance, small particles, with low lipophilicity and low amount of surface charges are preferable; 6) stability in biological fluids – once administrated, the radiopharmaceutical has to remain stable in a pH range of 1.5 to 8, and at 37ºC. Low stability could lead to an uptake in non-target tissues; 7) radiochemical purity and stability – the radiolabeled group must be stable against hydrolysis under physiological conditions, needs to have a strong affinity for the radionuclide and needs to be resistant to *in vivo* dehalogenation or transchelation. Otherwise, after radionuclide detachment, the imaging will no longer reflect the radiopharmaceutical fate (Saha, 2010; Gijs et al., 2016).

1.2. Imaging Techniques

Radionuclides molecular imaging techniques, including positron emission tomography (PET) and single-photon emission computed tomography (SPECT), have greatly contributed to the *in vivo* characterization of several diseases, providing information about a specific biological process, and improving the accuracy of detection, localization, and quantification of a disorder. The major advantages of SPECT and PET molecular imaging techniques are high sensitivity and specificity, allowing an accurate quantification, in addition to unlimited tissue penetration (Hong et al., 2011; Cuocolo et al., 2018).

In the interpretation of molecular images, is often common the use of a complementary imaging technique, such as computed tomography (CT) or magnetic resonance (MR). In PET and SPECT imaging, *in vivo* measurement of organ physiology, cellular metabolism, and perfusion is performed. However, these studies have poor resolution and lack anatomical detail. On the other hand, CT or MR imaging provides excellent spatial resolution with high anatomical detail, helping in accurate localization of lesions (Saha, 2006). Therefore, nowadays, most of the imaging machines are composed of a hybrid system in order to confer high sensitivity along with great spatial resolution.

1.2.1. Single Photon Emission Computed Tomography (SPECT)

The principle of single photon emission computed tomography (SPECT) has been well understood for many years and several medical installations were using SPECT since the early 1960s. A gamma camera acquires multiple planar views of the radioactivity in an organ, then, the data are processed mathematically to create cross-sectional views of the organ, creating a 3D image. SPECT utilizes the single photons emitted by gamma-emitting radionuclides, with energies between 100 to 300 KeV, such as ^{99m}Tc, ^{67}Ga, ^{111}In, and ^{123}I (Hong et al., 2011; Pawsner et al., 2006). Nowadays, SPECT technique is used to diagnose several diseases, including primarily endocrine and neuroendocrine disorders, infection/inflammation, benign and malignant bone diseases, radio-guided surgery, pulmonary circulation, coronary artery disease, and neurodegenerative disorders (Mariani et al., 2010).

1.2.2. Positron Emission Tomography (PET)

The first human PET was built in 1974 by Michael Phelps and Ed Hoffman at Washington University, but only in 1997, it has reached out into clinics, the year when the 2-deoxy-2-fluorine-18-fluoro-D-glucose (^{18}FDG), the most famous radiotracer used for cancer imaging, was approved, by the Food and Drug Administration (FDA), as a radiopharmaceutical (Nutt et al., 2002). PET imaging is based on the detection in the coincidence of gamma-ray emission originated from β^+ emitting sources. The main used isotopes are ^{11}C, ^{13}N, ^{15}O, ^{18}F, and ^{68}Ga. Positrons are annihilated in body tissue, producing two 511-keV photons that are emitted in opposite directions (180°). PET cameras are designed to detect the paired 511-keV photons generated from the annihilation event. Following the acquisition, the data are reconstructed in a manner similar to that used for SPECT in order to generate 3D images. PET has some advantages compared to SPECT, like its greater sensitivity and resolution and the existence of positron emitting isotopes for elements of low atomic number. This permits incorporation of positron emitters into many biologically active compounds, including isotopic forms of oxygen, carbon, nitrogen, and fluorine. However, the main disadvantage of PET is the high cost of the equipment and the short half-life of some of the most useful positron emitters. Currently, there are some PET radiopharmaceuticals available for tumor localization, cerebral and myocardial blood flow, volume and glucose metabolism (Phelps et al., 2000; Sharp et al., 2005; Saha, 2006; Pawsener et al., 2006). The main characteristics of both molecular imaging techniques are summarized in Table 1.

Table 1. Summary characteristics of PET and SPECT imaging

	PET	SPECT
Radionuclides	^{11}C, ^{18}F, ^{13}N, ^{68}Ga, ^{64}Cu	^{99m}Tc, ^{123}I, ^{67}Ga, ^{111}In
Energy (KeV)	511	100 - 300
Spatial resolution (mm)	1 -2	0.3 - 1
Sensitivity (mol/l)	$10^{-11} – 10^{-12}$	10^{-10} - 10^{-11}

PET: Positron Emission Tomography; SPECT: Single Photon Emission Computed Tomography.

1.3. Radionuclides

The main radioisotopes used for diagnosis purposes, nowadays, are technetium-99m (^{99m}Tc), iodine-123 (^{123}I), indium-111 (^{111}In), for SPECT imaging, and gallium-68 (^{68}Ga), copper-64 (^{64}Cu), and fluorine-18 (^{18}F) for

PET technique (Farzin et al., 2016; Muller et al., 2018). Each radionuclide has a specific physical half-life, decay mode, chemical properties, and a production method. Therefore, it is important to consider the characteristics of radionuclides to target the biological process or disease, which is to be visualized, characterized, or measured (Hassanzadeh et al., 2018). Table 2 shows the properties of prominent radionuclides, all of which are diagnostic radionuclides and emit a positron or gamma photon.

Table 2. Characteristics of common nuclear medicine radionuclides used in diagnosis purposes

Radioisotopes	Physical half-life	Production method	Decay mode	Imaging
^{99m}Tc	6.02 h	$^{99}Mo/^{99m}Tc$ Generator	IT	SPECT
^{123}I	13 h	Cyclotron	EC	SPECT
^{111}In	67 h	Cyclotron	EC	SPECT
^{68}Ga	68 min	$^{68}Ge/^{68}Ga$ Generator	β^+, EC	PET
^{64}Cu	12.7 h	Cyclotron	β^+, EC	PET
^{18}F	109.7 min	Cyclotron	β^+	PET

β^+ - Positron, EC - electron capture, IT - isometric transition.

1.4. Production of Radionuclides

There are three basic types of equipment that are used to make medical nuclides: cyclotrons, nuclear reactors, and generators:

1.4.1. Cyclotrons

Cyclotrons are circular devices in which charged particles such as protons, deuterons, and alpha particles are accelerated in a spiral path within a vacuum. They spiral outward under the influence of the magnetic field until they have sufficient velocity and are deflected into a target. When the charged particles interact with the targets of stable elements, nuclear reactions take place. A nucleus with excitation energy is formed and the excitation energy is disposed of by the emission of nucleons (i.e., protons and neutrons). Particle emission is followed by γ-ray emission when the former is no longer energetically feasible. Cyclotron-produced radionuclides are usually neutron deficient and therefore decay by β+ emission or electron capture. The main radionuclides produced in a cyclotron are ^{67}Ga, ^{111}In, ^{11}C, ^{13}N, ^{15}O, ^{18}F, ^{64}Cu. There are medical cyclotrons used to produce routinely short-lived radionuclides, particularly those used in positron emission tomography. These units are available commercially and can be installed in a relatively small space.

1.4.2. Nuclear Reactors

A general reactor is composed of fuel rods that contain large atoms (typically uranium-235, uranium-238, or plutonium-239) that are inherently unstable. These atoms undergo fission and two or three neutrons followed by approximately 200 MeV of heat energy are emitted during this process. Medical nuclides are made in reactors by the processes of fission or neutron capture. The main radionuclides produced in a reactor are ^{131}I, ^{67}Cu, ^{186}Re, ^{153}Sm and ^{99}Mo.

1.4.3. Generators

Generators are units that contain a radioactive "parent" nuclide with a relatively long half-life that decays to a short-lived "daughter" nuclide. The most commonly used generator is the technetium-99m (^{99m}Tc) generator which consists of a heavily shielded alumina column bounded with molybdenum-99 (^{99}Mo; parent). The ^{99m}Tc (daughter) is eluted by drawing sterile saline through the column into the vacuum vial. Others radionuclides are also produced in a generator, such as ^{68}Ga, ^{90}Y and ^{188}Re (Oliveira et al., 2006; Saha, 2006; Powsner et al., 2006; Saha, 2010).

1.5. General Strategies for Radiopharmaceuticals Radiolabeling

In a radiolabeled compound, similar or different radioactive atoms are added or substituted in the molecule. In any labeling process, a variety of physicochemical conditions can be employed to achieve a specific kind of labeling. There are mainly four methods employed in the preparation of labeled compounds for clinical use and they are summarized in Table 3.

The best method of radiolabeling will depend on the characteristics of the molecule of interest and the radionuclide of choice.

Table 3. General radiolabeling methods

Method	Concept	Example
Isotope exchange	One or more atoms in a molecule are replaced by isotopes of the same element having different mass numbers.	^{125}I-triiodothyronine (T3), ^{125}I-thyroxine (T4),
Introduction of a foreign label	A radionuclide is incorporated into a molecule primarily by the formation of covalent or coordinate covalent bonds	^{99m}Tc-radiopharmaceuticals
A compound to be labeled is attached to a molecule	The labeling is perfomed by a bifunctional chelating agent, a prosthetic group or other, added to the molecule of interest.	^{111}In-DTPA-albumin, ^{99m}Tc-DTPA-antibody, ^{64}Cu-NOTA-aptamer.
Biosynthesis	A living organism is grown in a culture medium containing the radioactive tracer, which is incorporated into metabolites produced by the organism.	^{57}Co cyanocobalamin, ^{14}C labeled compounds.

Table 4. Main bifunctional chelator groups (BFC) used in radiolabeling methods

BFC	NAME	FUNCTIONAL GROUP	ISOTOPE
MAG3	mercaptoacetyl-glycine-glycine-glycine	N-Hydroxyl succinimide	^{99m}Tc
HYNIC	hydrazinonicotinamide	Amine	^{99m}Tc
DTPA	diethylenetriaminepentaacetic acid	Carboxymethyl	^{99m}Tc, ^{111}In, ^{68}Ga
DOTA	1,4,7,10-tetraazacyclododecane-1,4,7,10-tetraacetic acid	Carboxymethyl	^{64}Cu, ^{68}Ga
NOTA	1,4,7-triazacyclononane-1,4,7-triacetic acid	Carboxymethyl	^{64}Cu, ^{68}Ga

Radiolabeling with radiohalogens (^{18}F, ^{76}Br, ^{125}I and others) is most often accompanied by the use of prosthetic groups, which are usually radiolabeled before conjugation with the targeting molecule. By the other side, radiolabeling with radiometals (^{64}Cu, ^{68}Ga, ^{111}In, ^{99m}Tc and others) is often performed by the use of bifunctional chelator groups (BFC), agents that possess at least two functional groups containing donor atoms capable of combining with a metal ion via a ring formation process, and with the molecule of interest (indirect labeling). The main BFC used in radiolabeling methods are summarized in Table 4.

Direct radiolabeling without the use of chelators have also been demonstrated mainly for ^{99m}Tc and ^{188}Re.

1.6. Targeting Molecules of Radiopharmaceuticals

In the clinics, most of the radiopharmaceuticals used are composed of synthetic molecules complexed with ^{99m}Tc. These radiopharmaceuticals are called "first generation" or perfusion radiopharmaceuticals. These types of complexes are transported across the blood and reach the target organ in proportion to blood flow. They do not have specific binding sites and they are probably distributed in the organism according to their size and surface charge. Nowadays, perfusion radiopharmaceuticals have important applications, including evaluation of cardiac (^{99m}Tc-sestamibi), brain (^{99m}Tc-ECD, ^{99m}Tc-HMPAO), pulmonary (^{99m}Tc-MAA), renal (^{99m}Tc-DTPA, ^{99m}Tc-DMSA), and bone (^{99m}Tc-MDP) function (Oliveira, 2006).

Nevertheless, many efforts have been made in order to develop specific radiopharmaceuticals, namely "second generation," which are specifically designed to target biological molecules, being able to bind to cellular receptors or be transported into specific cells. The most used classes of molecules are antibodies, peptides, nanoparticles, and, more recently, aptamers.

The use of antibodies as molecular target is mostly due to its high degree of specificity for binding to their target epitopes in a complex biological environment. Moreover, advances in the synthesis and biochemical engineering have been made in order to facilitate its use. An antibody, intact or fragment, could be used to diagnosis and/or therapy purposes. An intact antibody benefits from their molecular weight (~150 kDa) and size (~20 nm) avoiding fast renal clearance, while being conferred greater tumor bioavailability due to FcRn - mediated recycling in the systemic circulation. On the other hand, antibody fragments show better pharmacokinetics and also could have a good tumor targeting. For a radiopharmaceutical, mainly for diagnosis, a faster blood clearance is preferable, since a prolonged residence of the radiopharmaceutical in systemic circulation could impart unnecessary radiation dose to non-target organs. The most used radionuclides for antibody labeling are ^{18}F, ^{64}Cu, ^{68}Ga, ^{86}Y, ^{89}Zr, ^{99m}Tc, ^{111}In (Carter, 2017).

The fast advancing field of nanotechnology has generated several innovative drug delivery systems, such as liposomes, dendrimers, quantum dots, iron oxide, and carbon nanotubes, to improve and enhance targeted transport of cytotoxic drugs and radionuclides mainly to tumor areas. Nanosystems can reach the interest area by passive targeting - through the enhanced permeability and retention (EPR) effect in leaky tumor tissues, or active targeting – the nanocarrier has a surface modified group, such as specific antibodies or peptides, to actively targeted specific tumor or tissues. There are some approaches generally used for labeling nanocarriers: the radionuclide could be encapsulated into the nanosystem during preparation; the radionuclide could interact directly with the nanocarrier surface after preparation and the nanocarrier could have a bioconjugate in their surface that allows the complexation with the radionuclide. The most used radionuclides for nanotargeted imaging are ^{67}Ga, ^{111}In, ^{123}I, ^{99m}Tc, ^{64}Cu, ^{18}F (Ting, 2009).

Peptides are low molecular weight amino acid polymers (usually less than 10,000 kDa, and less than 100 amino acids). The majority of the peptides used in nuclear medicine are constituted by a relatively small number of amino acids (up to 30). Such small peptides do not show a well-defined tertiary structure. In contrast to bigger proteins and antibodies, peptides can be easily chemically synthesized, modified, and stabilized to obtain optimized pharmacokinetic parameters. Additionally, they are rapidly taken up and retained in target tissues, in accordance with the usually rapid plasma clearance due to the renal excretion, are not immunogenic and have high receptor binding affinity. Peptides can be synthesized by solid phase methods, which allows inserting directly a metal chelator, with or without a spacer, into the peptide facilitating the labeling procedure. Different radioisotopes are currently used for peptide labeling, such as ^{131}I, ^{99m}Tc,

^{111}In, ^{68}Ga, ^{64}Cu, ^{18}F. There are several peptides under study, for different application, such as in oncology, neurology, cardiology, inflammation, and infection diagnosis (Signore, 2001).

2. APTAMERS

Aptamers are oligonucleotides or peptides that bind to a specific target molecule. The term 'aptamer' is derived from the Latin word *aptus* (meaning ′to fit`) and the Greek word *meros* (meaning 'part'). The acid nucleic aptamers, which are the focus of the present chapter, are typically small single-strand synthetic DNA or RNA oligonucleotides (<100-mer, 10 to 25 kDa) with a unique three-dimensional structure that enable them to bind to a defined molecular target with high affinity and specificity.

Acid nucleic aptamers are generally developed *in vitro* by a molecular evolution process based on iterative selection-amplification steps known as Systematic Evolution of Ligands by Exponential Enrichment (SELEX), first introduced in 1990 (Tuerk and Gold, 1990). In a typical SELEX procedure a chemically synthesized random-sequence library (10^{14}-10^{15} sequences) constituted of DNA or RNA oligonucleotides flanked on each side by a known primer sequence is used. The fixed primer sequence at either side end allows the PCR amplification. The middle oligonucleotide region consisting of random nucleotides vary from 10 to 100 bases. This methodology includes the exposure of the target molecule to the library to allow the interaction of all binding oligonucleotides to the target. The next step involves separating the non-bounded sequences from the sequences that bind to the target for further PCR amplification (DNA aptamers) or RT-PCR followed of *in vitro* transcription (RNA aptamers). The enriched pool of aptamers is then exposed to the target again. The selection, separation and amplification protocols comprise a SELEX round. Usually, six to as many of 15 rounds have been reported to allow the best binders selection. After the aptamer selection process, the PCR products of the last round are cloned into a vector and sequenced for identification of the binding sequences. Finally, these sequences can be chemically synthesized.

Over the past three decades, more than 32 SELEX variations have been reported in an attempt to either reduce processing time, generate aptamers with novel designs and functions, or increase the process throughput (Wang et al., 2019). One of these variations is the cell-based SELEX methodology (cell-SELEX), which selects aptamer against a whole cell (Sefah et al., 2010). The cell-SELEX takes advantage that molecular targets on the cell surface are in their native state representing their natural folding structures. Aptamers have been developed by cell-SELEX for a wide variety of live cells, especially for cancer cells. To generate aptamers that can specifically target cancer cells, for example, the protocol includes negative selection steps to remove sequences binding to normal cells. Another interesting SELEX variation uses capillary electrophoresis (CE) for the separation between target-bound and unbound oligonucleotides (Mosing et al., 2009). The CE-SELEX protocol is capable of isolating high-affinity aptamers in as a little 2-4 rounds of selection.

Aptamers exhibit high binding affinity, having dissociation constants typically in the nM and even picomolar (pM) range (Thiviyanathan, V. & Gorenstein, D. G., 2012). They can provide high specificity by discriminating among closely related targets, such D and L amino acids (Geiger et al., 1996), a guanidine group and urea (Famulok, 1994), and the presence or absence of a methyl group (Jenilson et al., 1994) or a hydroxyl group (Sassanfar et al., 1993) in a molecule.

The wide variety of targets for which aptamers can be selected highlights their potential. Aptamers can bind to many types of small structures, such as metal ions, hormones, antibiotics, AMP and ATP, dye molecules and pesticide residues. Aptamers can also bind to macromolecules, such as antibodies and antigens, viruses, cytochromes, neurotoxins, and interleukins (Cai et al., 2018). In addition, the cell-SELEX technology allowed the selection of many aptamers against the specific cell, including various cancer cells, virus-infected cells, and microorganisms (Davydova et al., 2016).

Aptamer–target interactions rely on the characteristics of the targets and the flexible nature of aptamers. Commonly, aptamers can form a wide variety of structural shapes, such as hairpins, bulges, pseudoknots, and

G-quadruplexes. Based on these conformations, the forces that mediate the aptamer–target binding mainly involve hydrogen bonding, electrostatic interactions, hydrophobic effect, π–π stacking and van der Waals contacts (Tan et al., 2016).

Aptamers are often called "chemical antibodies" since they can be used as an alternative to antibodies from many applications. However, in comparison to the antibodies, aptamers have a number of advantages such as higher temperature stability, easy labeling, simple artificial synthesis, fast target accumulation and shortened body clearance. Aptamers have a flexible structure and a size 20-fold smaller compared with antibodies, allowing better tissue penetration. They fill the gap between small peptides (0.5-2 kDa) and single chain antibody fragments (~27 kDa). They are non-toxic and lack immunogenicity being unlikely to elicit an immune response, at least at the antibody level. Because aptamers are isolated entirely *in vitro*, toxic compounds and molecules with little or no immune response can also serve as targets. Their production by chemical synthesis is more reliable and reproducible than antibodies production. Appropriated functional groups can be easily incorporated into the aptamer synthesis step to allow subsequent incorporation of bioconjugates (e.g., fluorescent tags, radiometal chelators, biotin and drugs) or to modify their biokinetics (Missailidis & Perkins, 2007).

Research conducted during the last years has provided evidence that aptamers are promising tools in nuclear medicine. The present chapter reviews the efforts made so far to explore the aptamers as new radiopharmaceuticals.

2.1. Aptamers Advantages and Drawbacks for Application as Radiopharmaceuticals

Because of their unique characteristics, aptamers are promising tools for radionuclide-based imaging. They show high binding affinity to molecular targets with high specificity, which makes them effective biomolecules for generating excellent molecular imaging probes. Because of their smaller size

(5 to 25 kDa) and hydrophilicity aptamers have the ability to circulate freely in the blood, penetrate tissues more readily, reach peak levels more rapidly and clear faster from the body. These properties allow obtaining high-resolution images with high signal-to-noise ratios and reduced radiotoxicity for the patient (Bouvier-Müller & Ducongé, 2018). Aptamers have the specificity of antibodies with the flexibility of small molecules.

Aptamers can be labeled with different radioisotopes such as ^{99m}Tc, ^{188}Re, ^{111}In, ^{18}F, ^{64}Cu, ^{68}Ga, among others, thus allowing its application for imaging, therapy or theranostic according to the radionuclide used. Chemical modifications that allow the attachment of radionuclides chelators or prosthetic groups can be made in a straightforward manner and it is worth mentioning in the literature that radiolabeling did not affect the aptamer-target binding affinity (Hassanzadeh et al., 2018). Aptamer immunogenicity has never been reported, which allows them to be administered repeatedly to the same individual, for diagnosis or therapy. They have no side effects in treatments they have already been used. Generally, the radiolabeled aptamers show a high renal clearance. The elimination through the renal rout is preferred for radiopharmaceuticals over the hepatobiliary clearance, which presents a slow gastrointestinal transit (Farzin et al., 2019). Lyophilized aptamers exhibit higher stability for storage; they can be stored for years or subject to many freeze-thaw cycles without loss of activity (Pendergrast et al., 2005). These features contribute to the use of aptamers as components in lyophilized kits for fast radiolabeling.

However, aptamers have limitations for use as radiopharmaceuticals for diagnosis or therapy. The most important is susceptibility to degradation by nucleases in biological environments and, for therapy application, the rapid elimination by the renal route.

The half-life in the plasma of the oligonucleotide RNA is a few minutes, whereas for DNA it is around 30 to 60 min (White et al., 2000). Fortunately, aptamers can be easily modified to extend their *in vivo* half-life. Several chemical modifications have been introduced to protect them from nuclease degradation including modified nucleotides, changes on the ends of the oligonucleotide sequence and modifications on the phosphodiester linkage. Modified nucleotides can be incorporated during enzymatic synthesis using specialized DNA or RNA polymerases. However, the options are limited since not all modified nucleotides are compatible with the enzymatic steps of SELEX. Modified nucleotides can also be incorporated during the chemical solid-phase synthesis after selection. Post-selection modifications are cheaper and easier but can modify the aptamer binding capacity, which needs to be re-evaluated. Modifications to the sugars as 2'-amino (2'NH_2) pyrimidines, 2'-fluoro (2'F) pyrimidines or 2'O-methyl (2'OMe) ribose purines have been frequently used. Nuclease stabilization using locked nucleic acids, which covalently bridge the 2' and 4' ribose positions (Lipi & Chen, 2017) has also been explored. Another modification that is often used is the inclusion of inverted thymidines as terminal caps at 3' and/or 5' end of the aptamer to prevent the degradation by exonucleases. Modifications at the 3' and/or 5' end can increase the aptamer half-life from minutes to hours (Khan & Missailidis, 2008). Stabilizing backbone modifications are incorporated by replacing phosphor by sulfur in the internucleotide linkage. Replacement of the oligonucleotide phosphodiester linkage with triazole linkages is also effective.

However, not all aptamers show high nuclease vulnerability and this question has to be evaluated case by case. For some DNA aptamers, the three-dimensional folding can protect it by hiding specific target sequences for the action of endonucleases or the ends of the sequence (mainly the 3'end) that can be attacked by exonucleases.

Aptamers are easily cleared by the kidneys. Ninety percent of a typical aptamer administrated intravenously can wash out in the first 15 min (Healy et al., 2004). Whilst this behavior is interesting for diagnostic applications is disadvantageous for most therapy applications since the radiotracer will not be able to accumulate properly in the target. One way of solving this problem is to increase the molecular weight of the aptamer above the molecular cutoff for the renal glomerulus (50 kDa) by conjugation with polyethylene glycol, cholesterol, biotin-streptavidin and liposomes or formation of multimers, as tetrameric aptamer conjugates (Gijs et al., 2016).

The negative charge of aptamers may favor plasma proteins binding since most of them are positively charged (Henry et al., 2012). Some studies have suggested that the reversible binding of aptamers to plasma proteins seems to be helpful for imaging since it prevents very fast aptamer clearance by glomerular filtration (Gijs et al., 2016; Ferreira et al., 2017; Santos et al., 2017).

The applicability of aptamers-based radiopharmaceuticals is also reduced for targets in the brain tissue due to the blood-brain-barrier (BBB) that can be crossed only by small lipophilic molecules (<400 Da). To address this issue, aptamer conjugation to a brain targeting system to cross BBB, like the TGN peptide, was reported (Gao et al., 2012).

2.2. Aptamers Radiolabeling

As discussed before, two types of imaging techniques are employed in nuclear medicine, (1) SPECT that employ γ-emitting radionuclides, mainly ^{99m}Tc, ^{67}Ga, and ^{111}In, and (2) PET, which employ β^+-emitting radionuclides including ^{18}F, ^{68}Ga, ^{64}Cu, among others. Some radionuclides as ^{99m}Tc, ^{18}F, ^{64}Cu, ^{67}Ga and ^{68}Ga are among the most suited for aptamer radiolabeling as their physical half-lives' are similar to the aptamers biological half-life.

2.2.1. ^{99m}Tc

Currently, almost 80% of radiopharmaceuticals are based on ^{99m}Tc due to its adequate physical characteristics and chemical properties as the half-life of 6.02 h and pure gamma emission with the energy of 140,51 KeV. It is readily available from Mo generator as a product of ^{99}Mo decay. Technetium is a transition metal of silvery gray color belonging to group VIIB (Mn, Tc, and Re) and has the atomic number 43. It can exist in eight oxidation states, namely, 1- to 7+, being 7+ and 4+ the most stable states. The coordination number of ^{99m}Tc-complexes can vary between 4 and 9, being the stability of these states depending on the type, number of ligands, and chemical environment. The chemical form of ^{99m}Tc obtained from the generator is sodium pertechnetate (^{99m}Tc-$NaTcO_4$), which is a rather nonreactive species and does not label any compound by direct addition. Therefore, the use of reducing agents, to reduce ^{99m}Tc from the 7+ state to a lower oxidation state, is necessary. Stannous chloride is the commonly used reducing agent in most preparations, but stannous citrate, stannous tartrate and borohydride ($NaBH_4$) are also options. The reduced ^{99m}Tc species are chemically reactive and combine with a wide variety of chelating agents. The chelating agent usually donates pairs of electrons to form coordinate covalent bonds with reduced ^{99m}Tc. Chemical groups such as –COOH, –OH, $–NH_2$, and –SH can be used as the electron donors. The bounding with ^{99m}Tc may occur between the donor groups from the molecule itself (direct labeling) or through a bifunctional chelator groups (BFC), agents that possess at least two functional groups containing donor atoms capable of combining with a metal ion via a ring formation process, and with the molecule of interest (indirect labeling). Labeling procedures have been greatly facilitated by kit preparations. Sterile kits for labeling containing the chemical ingredients in the lyophilized form are commercially available and used to prepare ^{99m}Tc pharmaceuticals shortly before application to the patient (Zolle et al., 2006; Saha et al., 2010).

Aptamers radiolabeling with ^{99m}Tc is a challenging process that usually requires additional steps of synthesis and purification in order to conjugate the aptamer with the chelating molecule for the radionuclide. Some chelating approaches have been used successfully to stabilize the radioactive core, mainly Mercaptoacetildiglycine (MAG_2), MAG_3, MetCyc, HYNIC, DTPA and DOTA (Farzin et al., 2019). The incorporation of a chelator to the aptamer is usually performed prior to radiolabeling. The chelators useful with ^{99m}Tc can also form a stable complex with ^{188}Re because of similarities in the chemistry of technetium and rhenium. Radiolabeling with ^{99m}Tc can also be performed directly without the need to add a chelator. Correa et al., (2014) reported a method for direct aptamer radiolabeling with ^{99m}Tc that allows high radiochemical yield showing *in vitro* stability in the presence of plasma and cysteine. Tricine and Ethylenediamine-N,N'–diacetic acid (EDDA) were used as co-ligands. This method permits easy and rapid preparation, less cost and manipulation thus decreasing the chances of aptamer degradation.

2.2.2. ^{67}Ga and ^{68}G

The gamma emission of ^{67}Ga is imaged by a SPECT while the positron emission of ^{68}Ga is imaged by PET. The ^{67}Ga (half-life 3.3 days) is cyclotron produced. The ^{68}Ga (half-life 68 min) is mainly obtained from a ^{68}Ge/^{68}Ga generator, which is manufactured in cyclotron and commercially available in nuclear medicine centers. In aqueous solution, gallium occurs solely in the oxidation state 3+. Ga^{3+} is classified as a hard acid metal, bonding to highly ionic hard base ligand donors, such as carboxylic acids, amino nitrogens, hydroxamates, phenolates, and thiols groups. Like the ^{99m}Tc, these radiometals are most often incorporated in aptamers by the covalent attachment of a bifunctional chelator. Several suitable bifunctional chelators were proposed, developed, and coupled to biomolecules for gallium labeling, such as desferal, DOTA, NOTA, NODAGA, and NODAGATOC. Due to the high stability of Ga^{3+} - NOTA complexes, this chelator has been mostly used for the development of ^{67}Ga and ^{68}Ga based radiopharmaceuticals. The NOTA radiolabeling can be achieved at room temperature under mild conditions compatible with the handling of aptamers (Price & Orvig, 2014).

2.2.3. ^{64}Cu

The ^{64}Cu (half-life 12.7 h) is also a radiometal requiring the attachment of a bifunctional chelator to the aptamer before the radiolabeling (Willian & Rockey, 2011). The radionuclide can be produced with high specific activity by a medical cyclotron. The ^{64}Cu displays both β-minus decay and positron emission with applications for PET imaging and targeted radiotherapy of cancer. The well-established coordination chemistry of copper allows its reaction with a wide variety of chelator systems that can be linked to aptamers. The most used method to radiolabel compounds with ^{64}Cu is through macrocyclic chelators (DOTA, NOTA, TETA, TE2A, BAT) since simpler chelator agents (DTPA, DTTA, HYNIC) do not bind copper stably under physiological conditions. DOTA has been employed to label oligonucleotides with ^{64}Cu under chemically mild conditions, but it has been shown to be prone to demetallation *in vivo* due to transchelation with liver proteins. The NOTA is proved to be more suitable than the DOTA as the coordination number of Cu^{2+} is better matched (Liu et al., 2009). A promising chelator for Cu radioisotopes is 2,2'-(1,4,8,11-Tetraazabicyclo[6.6.2]hexadecane-4,11-diyl)diacetic acid (CB-TEA2) that forms very stable complexes and shows improved background clearance (Wei et al., 2007), even though it requires temperatures close to 90°C for radiometal incorporation.

2.2.4. ^{18}F

The leading radionuclide for PET imaging is the ^{18}F (half-life 110 min). Aptamer radiolabeling with radiohalogens like ^{18}F is usually performed by the use of prosthetic groups (also referred as ^{18}F-labeling building blocks), which formation requires many and complex reactions. Radiolabeling of prosthetic groups is frequently performed before conjugation with the aptamer due the harsh non-aqueous reaction conditions needed. Some of the approaches used to produce ^{18}F labeled aptamers are described below.

Click chemistry is the most used method to radiolabel compounds with ^{18}F, and it is a generic term for a set of labeling reactions, which make use of several selective and modular building blocks and enable chemo-selective ligations to radiolabel biologically relevant compounds. In most of the cases, radiolabeling with fluorine-18 is achieved by the bioisosteric exchange of single hydrogen or OH groups. This is due to the fact that fluorine is absent in most biologically active compounds. By click chemistry, 5' end alkynyl functionalized aptamers can be labeled with ^{18}F-benzylazide, which can be obtained by an automated radiochemical synthesis using spirocyclic hypervalent iodine(III) precursor for radiofluorination with [^{18}F]fluoride (Rotstein et al., 2014).

Aptamers with a hexylamine linker at 5' position can also be radiolabeled using ([^{18}F]fluoromethyl)phenyl isothiocyanate (De Vries & Vroegh, 2003), that can be synthesized by nucleophilic substitution using [^{18}F] fluoride and isothiocyanates substrates containing leaving groups.

Another method uses 2,5-dioxopyrrolidin-1-yl 4-(fluoro-^{18}F)benzoate (^{18}F-SFB) to react with aptamers functionalized with a primary amine at the 5' position, usually spaced by a six carbon chain (Orit & Xuefeng, 2015). The conjugation of ^{18}F-SFB to the aptamer is achieved by a reaction performed in a basic sodium phosphate buffer, which presents a low but usable yield.

Hybridization-based ^{18}F-radiolabeling is another useful process in which an ^{18}F pre-labeled oligonucleotide is hybridized with a complementary sequence in the aptamer (Park & Lee, 2016).

3. Overview of *In Vivo* Studies Using Aptamers as Radiopharmaceuticals

The use of aptamers as radiopharmaceuticals is relatively new. However, several aptamers have been tried for this purpose and the number of reports has increased significantly in recent years. The vast majority was addressed to cancer and all of them tested only at the preclinical level. In this section, will be presented in a review format an overview of the studies performed with radiolabeled aptamers focused on those that reached the *in vivo* evaluation stage (Table 5).

The first study to test aptamers as *in vivo* imaging probes was the use of the DNA aptamer NX21909 that binds to neutrophil elastase. This enzyme is released by neutrophils accumulated at the site of inflammation. The aptamer was directly radiolabeled with ^{99m}Tc by a method using TCEP as reducing agent. The radiolabeled aptamer was evaluated by scintigraphy for diagnostic imaging of inflammation in a rat model of the reverse passive Arthus reaction induced by immune complex deposition in a forelimb (Charton et al., 1997).

The radiolabeled aptamer biodistribution was compared to that of a reference ^{99m}Tc-IgG used to image inflammation and a negative control ^{99m}Tc-oligonucleotide. A higher signal was measured with ^{99m}Tc-NX21909 than the control ^{99m}Tc-oligonucleotide in the inflamed forelimb. The radiolabeled aptamer showed faster clearance than the ^{99m}Tc-IgG leading to a signal-to-noise ratio of 4.3 just 2 h post-infection, whereas the best ratio for the ^{99m}Tc-IgG was of 3.1 obtained 3 h post-injection. These results indicated that the aptamer technology for diagnostic imaging might be superior over antibodies.

Table 5. Aptamers that have been radiolabeled for *in vivo* studies

Name	Target	Chemistry	Labeling	Radiolabeling Strategy	Potential use	References
TTA1	Tenascin C	2'OMe-Pu, 2' F-Py RNA	^{99m}Tc	Via chelator (MAG2, MAG2-PEG_{3400} and DTPA)	Cancer imaging	Hicke et al., 2006.
TTA1	Tenascin C	2'OMe-Pu, 2' F-Py RNA	^{18}F	Via prosthetic group (^{18}F-FPyME)	Cancer imaging	Boisgard et al., 2009.
GBI-10	Tenascin C	DNA	^{18}F or ^{64}Cu	Via prosthetic group (^{18}F-SFB) and chelator (^{64}Cu-NOTA)	Cancer imaging	Jacobson et al., 2015.
AptaA	Mucin 1	DNA	^{99m}Tc	Via chelator (MetCys, MAG_3, DOTA and carboxy-porphyrin)	Cancer imaging	Borbas et al., 2007.
AptaA and AptaB	Mucin 1	DNA	^{99m}Tc	Via chelator (MAG_2)	Cancer imaging	Da Pieve et al., 2009.
AptaA	Mucin 1	DNA-PEG	^{99m}Tc	Via chelator (MAG_2)	Cancer imaging	Pieve et al., 2012.
AptaA and AptaB	Mucin 1	DNA	^{99m}Tc	Via nanoparticles (mesoporous silica)	Cancer imaging and drug delivery	Sa et al., 2013; Pascual et al., 2017.
AptaA and AptaB	Mucin 1	DNA	^{99m}Tc	Via nanoparticles (Poly(latic-*co*-glicolic acid))	Cancer imaging	Carmo et al., 2017.
uMUC-1	Mucin 1	DNA	^{68}Ga	Via chelator (NOTA) on multimodal nanoparticle.	Cancer imaging	Kang et al., 2015.
AS1411	Nucleolin	DNA	^{64}Cu	Via chelator (DOTA-NHS, CB-TE2A, DOTA-Bn, NOTA-Bn)	Cancer imaging	Li et al., 2014.
AS1411	Nucleolin	DNA	^{99m}Tc	Via chelator (HYNIC)	Cancer imaging	Noaparast et al., 2015.
AS1411	Nucleolin	DNA	^{68}Ga	Via chelator (NOTA) on multimodal nanoparticle	Cancer imaging	Hwang et al., 2010.
F3B	hMMP-9	2'OMe-Pu, 2' F-Py RNA	^{99m}Tc	Via chelator (MAG_3)	Cancer imaging	Da Rocha Gomes et al., 2012.

Table 5. (Continued)

Name	Target	Chemistry	Labeling	Radiolabeling Strategy	Potential use	References
F3B	hMMP-9	2'OMe-Pu, 2' F-Py RNA	^{99m}Tc or ^{111}In	Via chelator (MAG_3 and DOTA)	Cancer imaging	Kryza et al., 2016.
U2	EGFRvIII	DNA	^{188}Re	Direct radiolabeling	Cancer imaging	Wu et al., 2014.
E07	EGFR	2' F-Py RNA	^{111}In	Labeling of gold nanospheres via chelator (DTPA)	Cancer imaging and therapy	Melacon et al., 2014.
ME07	EGFR	2' F-Py RNA	^{18}F	Via prosthetic group (^{18}F-flucrobenzoyl)	Cancer imaging	Cheng et al., 2015.
HER2	HER2	2' F-Py RNA	^{99m}Tc	Via chelator (HYNIC)	Cancer imaging	Varmira et al., 2013; Varmira et al., 2014.
(HeA2_3 and HeA2_1	HER2	DNA	^{68}Ga	Via chelator (NOTA)	Cancer imaging	Gijs et al., 2016.
SH1194-35	HER2	DNA	^{18}F	Via prosthetic group (^{18}F-SFB)	Cancer imaging	Kin et al., 2019.
Sgc8	PTK7	DNA	^{18}F	Via prosthetic group (azido ^{18}F-arenes)	Cancer imaging	Wang et al., 2015.
Sgc8-c	PTK7	DNA	^{99m}T	Via chelator (HYNIC)	Cancer imaging	Calzada et al., 2017.
NX21909	Neutrophil elastase	DNA	^{99m}T	Direct radiolabeling	Inflammation imaging	Charton et al., 1997.
ODN1 & ODN2	Thrombin	DNA	^{123}I	Via solid-phase synthesis	Thrombus imaging	Dougan et al., 2003.
SA20, SA23 and SA34	*S. aureus*	DNA	^{99m}Tc	Direct radiolabeling	Infection imaging	Santos et al., 2015; Santos et al., 2017.
Antibac1	Peptidoglycan	DNA	^{99m}Tc	Direct radiolabeling	Infection imaging	Ferreira et al., 2017.

An attempt of thrombus-imaging was performed by Dougan et al., (2003). The DNA aptamer ODN2 that reacts with the heparin-binding site on thrombin (exocite 2) was tested *in vivo* in a rabbit jugular vein thrombosis model. Thrombin is the last enzyme in the clotting cascade cleaving fibrinogen to fibrin. The ODN2 aptamer was labeled with ^{123}I and inoculated via the truncated facial vein. The thrombus area was imaged by scintigraphy, or excised for counting. The results showed that the ^{123}I-ODN2 uptake at the thrombus area was equal to the ^{125}I-ovalbumin (negative control). Furthermore, thrombin-dependent images could not be obtained. The rapid aptamer clearance from blood, combined with slow mass transfer in the thrombus appears to be the cause of this failure.

To date, the reported applications of aptamers as radiopharmaceuticals have been mostly focused on cancer and exploring few molecular targets. The radiolabeled RNA aptamer TTA1 was the first aptamer used as a molecular cancer-imaging agent. It is addressed against the protein tenascin-C (TN-C), an extracellular matrix protein that is overexpressed during tissue remodeling processes, including tumor growth. The TTA1 was generated by combining cell-SELEX on TN-C positive U251 glioblastoma cells with protein-SELEX against purified recombinant TN-C (Hicke et al., 2001). To improve stability against plasma nucleases, every pyrimidine of this aptamer (39-mer) contained a fluor group in the 2' position of the ribose. It also was modified with the addition of maximal 2'-OMe purine substitutions and an inverted thymidine cap at the 3' end. The TTA1 was conjugated to MAG_2, MAG_2-PEG_{3400} or DTPA via a hexyl-aminolinker at the 5' end and radiolabeled with ^{99m}Tc or ^{111}In. The biodistribution was measured *ex vivo* and by SPECT imaging in xenografts murine models of glioblastoma and breast cancer (Hicke et al., 2006). The biodistribution of ^{99m}Tc-MAG_2-TTA1 showed a maximum uptake of 6% of injected dose/g (ID/g) in the tumor 10 min after intravenous injection and it was of 1.9% 3 h post-injection. The uptakes of the negative control radiolabeled aptamer were of 3% and 0.04% at 10 min and 3 h, respectively. After 10 min post injection the tumor was faintly visible, prominent at 3 h and at 18 h it was clearly visualized as blazing structure. The biodistribution of ^{99m}Tc-MAG_2-TTA1 showed rapid renal and hepatic clearance reaching a tumor-to-blood ratio of 50 just 3 h post injection. Substitution of ^{99m}Tc-MAG_2 by DTPA-^{111}In increases dramatically hepatic uptake highlighting the radiometal chelator importance in the biodistribution profile. The PEGylated ^{99m}Tc-MAG_2-TTA1 was predominantly eliminated by renal clearance and subsequent excretion by the urine.

The aptamer TTA1 was also used in the first application of an aptamer for PET imaging (Boisgard et al., 2009). The tiol-functionalized aptamer was radiolabeled with ^{18}F using the pre-labeled 1-[3-(2-[^{18}F]-fluoropyridin-3-yloxy)propyl]pyrrole-2,5-dione (^{18}F-FPyME). Whole body biodistributions and pharmacokinetics of [^{18}F]TTA1 were evaluated in Wistar rats and in U251 human glioblastoma tumors-bearing nu/nu mice. In both species, the main routes of body elimination of [^{18}F]TTA1 were the urinary tract and the hepato-enteric pathway. Tumoral uptake was rapid with a maximum observed 15 min after injection, corresponding to 1.7% of ID/cc, and decreasing to 0.9% at 90 min. However, unspecific muscular uptake was high yielding a tumor-to-muscle ratio of 2.3 and 5.4 at 15 and 90 min, respectively.

Another aptamer addressed to TN-C was evaluated by PET imaging. The DNA aptamer GBI-10 was select by cell-SELEX using the glioblastoma-derived cell line U251 (Daniels, 2003) as target. It was radiolabeled with ^{18}F or ^{64}Cu, and tested in mice bearing a subcutaneous xenograft tumor (Jacobson et al., 2015). For labeling with ^{18}F, the aptamer 5’ end was functionalized with a primary amine spaced by a six carbon chain and then it was linked to the ^{18}F-SFB (Orit & Xuefeng, 2015). The labeling with ^{64}Cu was performed after the aptamer conjugation to the *p*-SCN-Bn-NOTA chelator. The radiolabeled GBI-10 aptamer, for both labeling strategies, provides a clear visualization of tenascin-C-positive tumors, but not tenascin-C-negative tumors. The radiolabeled GBI-10 aptamer uptake was also significantly higher than a non-specific scrambled aptamer (negative control) and it had fast clearance through the kidneys, resulting in high tumor contrast. However, the ^{99m}Tc-MAG_2-TTA1 aptamer used for SPECT in the study of Hicke et al., (2006) showed 10-fold higher uptake in the tumor 1 h after injection and 10-fold higher tumor/blood ratio. This could be explained by the fact that GBI-10 has lower affinity at 37°C

(Kd in the μM range) and it has not been modified to improve stability against plasma nucleases.

Mucin 1 glycoprotein (MUC1) is a large membrane-associated glycoprotein whose function is lubrication and hydration of cells surfaces as well as protection from microorganisms and degradative enzymes. It is a well-known tumor marker highly expressed by the majority of cancers and, in particular, by primary and metastatic breast cancers (Yousefi et al., 2019). MUC1 overexpression has been associated with poor clinical prognosis. The aptamers AptA and AptB were select against the MUC1 protein core or the glycosylated protein, respectively, by SELEX against purified MUC1 peptides (Ferreira et al., 2006). *In vivo* aptamers stability was improved by including an inverted thymidine at 3' end as well as 5'-amino group to enable additional conjugation. The MUC1 aptamers were labeled with ^{99m}Tc using MetCys or MAG_3 chelators and evaluated in mice bearing xenografted MCF7 breast cancer tumor. The ^{99m}Tc-MetCyc-AptaA presented higher tumor uptake due to very little liver and kidney uptake. By the other side, ^{99m}Tc-MAG_3-AptaA displayed fast renal clearance and high kidney and liver uptake. In an attempt to slow down renal clearance two tetrameric conjugates were synthesized by the attachment of four aptamers to a central chelator (DOTA or porphyrin ligand 4). Despite the tetrameric complexes showed improved tumor uptake and increased retention in the circulation compared to monomeric compounds, the biodistribution demonstrated free ^{99m}Tc in the stomach and large intestine evidencing that the labeling efficiency was lower (Borbas et al., 2007).

AptaA and AptaB have also been conjugated in high yield to MAG_2 and labelled with ^{99m}Tc. The biodistribution properties of the two radiolabeled aptamers were compared in MCF-7 tumor bearing mice (Pieve et al., 2009). Both aptamers show a maximum tumor uptake after 5 h post injection and AptB has a slightly higher accumulation rate, 0.12 %ID/g for AptA and “0.14 %ID/g for AptB. Most of the activity was cleared from the system by 3 h. The tumor uptake and clearance data suggested a quicker internalization of AptaB in relation to AptaA in the tumor cells. To overcome the rapid kidney clearance and to improve tumor uptake, the same research group increased the MW of the aptamer AptA by conjugation with polyethylene glycol (PEG). PEG 20 kDa or polyPEG 17 kDa were connected do 3' end of the aptamer by thiol-maleimide conjugation. MAG_2 was used as a chelator for ^{99m}Tc radiolabeling. Biodistribution was evaluated in MCF-7 tumor bearing mice. However, the increased molecular weight seems to decrease the infiltration of the complex in the tumor tissue since low tumor uptake was verified (Pieve et al., 2012).

The MUC1 aptamers were loaded on silica nanoparticles and the complex labeled directly with ^{99m}Tc in the presence of stannous chloride. The biodistribution in health mice pointed high uptake in the lungs (10 %ID/g) and renal and hepatic clearance (Sa et al., 2013). In another study, mesoporous silica nanoparticles functionalized with aminopropyl groups were associated with MUC1 aptamers (S1-apMUC1). These nanoparticles labeled with ^{99m}Tc showed a remarkable tumor accumulation in MDA-MB-231 tumor bearing mice. The tumor targeting was observed in biodistribution studies and SPECT images, 120 min and 90 min after intraocular injection, respectively. The uptake of the radiolabeled S1-apMUC1 nanoparticles in the tumor (20% ID/g) was significantly larger compared with the accumulation reported for free radiolabeled MUC1 aptamer. The uptake in lung and intestines was attributed to the presence of MUC1 glycoprotein in these organs, while low uptake was verified in the liver and spleen. The clearance was predominantly by the renal route. The results suggested that the ^{99m}Tc radiolabeled S1-apMUC1 is a potential candidate for cancer diagnosis by imaging (Pascual et al., 2017).

Poly(latic-*co*-glicolic acid) nanoparticles loaded with MUC1 aptamer and labeled with ^{99m}Tc were also used for biodistribution study and imaging of triple negative breast cancer (TNBC). TNBC is characterized by low expression of the estrogen receptor, progesterone receptor and human growth factor receptor 2, and it tends to be more aggressive than the others breast cancer subtypes. The biodistribution in an induced murine model with TNBC showed a high tumor uptake (5%) and a high capture by the intestine (>30%). The high uptake by the tumor was confirmed by SPECT imaging. The renal clearance and the low uptake by the liver and spleen suggested its possible application in human beings (Carmo et al., 2017).

An aptamer was selected to underglycosylated mucin-1 (uMUC-1) antigen, which is highly expressed by the majority of adenocarcinomas. Cobalt ferrite magnetic nanoparticles surrounded by fluorescent rhodamine within silica shell matrix were conjugated with the aptamer and further labeled with ^{68}Ga with the help of a *p*-SCN-bn-NOTA chelating agent, resulting in a multimodal nanoparticle. This multimodal nanoparticle can be used for optical, magnetic resonance (MR) and positron emission tomography (PET) imaging. Multimodal imaging is useful for overcoming the limitations of a single imaging modality. For *in vivo* multimodal cancer imaging, BT-20 cells (a human breast cancer cell line) were incorporated within a PPLA scaffold and implanted into the right thighs of nude mice. The left thighs were implanted with cell-free PPLA scaffold as control. The implanted cells were cultured for 3 weeks. Then, the multimodal nanoparticles (MFR-uMUC-1) or control nanoparticles loaded with a non-binding aptamer (MF-uMUC-1 mt) were injected by the tail vein. The uptake of ^{68}Ga radioactivity in the right thigh of MFR-uMUC-1 injected mice was demonstrated by dynamic small-animal PET image at 2 h after injection. The MR images 24 h after injection showed dark signals in the right thigh of MFR-uMUC-1 injected mice. No significant signals in fluorescence, radioactivity and MR intensity were verified in the left thighs or in the right thighs of mice inoculated with the control nanoparticles. The nanoparticles showed rapid clearance through the blood-stream. These results showed a high specificity for targeting cancers expressing uMUC-1 *in vivo* (Kang et al., 2015).

Nucleolin is a phosphoprotein abundant in the nucleolus. It is translocated to the external side of the cell membrane to serve as a receptor for many molecules involved in cell differentiation, adhesion, inflammation, angiogenesis and tumor development. Nucleolin levels on the cell surface are correlated with cell proliferation. This protein is overexpressed in tumors such as breast cancer, hepatocellular, lymphocytic leukemia, prostate and renal carcinoma tumors. The aptamer AS1411 (also known as AGRO1000) is a 26-mer guanine-rich oligonucleotide (GRO) that binds to nucleolin in the plasma membrane (Girvan et al., 2006). This aptamer has an anti-proliferative effect on tumor cells and it has completed a phase 2 trial

(Rosenberg et al., 2014). The PET isotope ^{64}Cu was used to radiolabel the AS1411 aptamer and the complex was explored as a diagnostic imaging agent for detecting lung cancer (Li et al., 2014). Four chelators were tried to label the aptamer, which affected the cellular uptake (*in vitro*) in the following order: DOTA-NHS > CB-TE2A > DOTA-Bn > NOTA-Bn. MicroPET imaging with ^{64}Cu-CB-TE2A-AS1411 showed clear tumor uptake 1 and 24 h post injection, whereas tumors were undetected for up 24 h with ^{64}Cu-NOTA-AS1411. The ^{64}Cu-CB-TE2A-AS1411 also demonstrated lower liver uptake and higher tumor-to-background ratio. The AS1411 aptamer was also evaluated to target prostate tumor cells. The aptamer was conjugated to HYNIC and labeled with ^{99m}Tc using tricine as co-ligand. Based on biodistribution assessment of ^{99m}Tc-HYNIC-AS1411, tumor uptake was 2.33% ID/g and 0.85% ID/g at 1 and 4 h, respectively. Rapid blood clearance was observed after injection and the route of elimination was the urinary system (Noaparast et al., 2015).

The AS1411 aptamer was used to construct a multimodal cancer-target nanoparticle capable of simultaneous *in vivo* fluorescence imaging, radionuclide imaging (SPECT) and magnetic resonance imaging in mice (Hwang et al., 2010). For this purpose, a cobalt-ferrite nanoparticle surrounded by fluorescent rhodamine within a silica shell matrix was synthesized with the AS1411 aptamer. After purification, the particle was bound to *p*-SCN-bn-NOTA chelating agent and labeled with ^{67}Ga. The multimodal nanoparticles were injected into mice xenografted with nucleolin-expressing C6 rat glioma cells. Scintigraphic images and magnetic resonance imaging showed specific targeting of cancer cells at 24 h after injection. However, high liver accumulation was verified.

The human Matrix MetalloProtease-9 (hMMP-9) enzyme is involved in tissue remodeling, cancer metastasis and arthritis. The enzyme contributes to tumor metastasis by releasing cancer cells and it is a relevant marker of a malignant tumor. The RNA aptamer F3 (68-mer) was selected to hMMP-9 with high affinity (Kd = 20 nM) and specificity. The aptamer was further truncated to F3B (36-mer) and modified to become nuclease resistant using 2'O-methyl purines and 2' fluoropyrimidines. The modified aptamer was conjugated to MAG_3 and radiolabeled with ^{99m}Tc. The radiolabeled aptamer was able to detect the presence of hMMP-9 in *ex vivo* imaging slices of human brain tumors (Da Rocha Gomes et al., 2012). The increased binding was visualized with the grade of malignancy and no signal was detected with healthy brain tissue. Slices of human brain tumors were used since murine xenograft model has some limitations for representing many aspects of human real tumor.

Kryza et al., (2016) performed quantitative biodistribution studies with ^{99m}Tc-MAG$_3$-F3B and ^{111}In-DOTA-F3B in A375 melanoma bearing mice. The ^{99m}Tc-MAG$_3$-F3B specifically detected hMMP-9 in melanoma tumors (1.8% ID/g) but accumulation in the digestive tract was very high (> 18% ID/g), limiting the detection of melanoma tumor and their metastases in the abdominal region. The digestive tract uptake of ^{111}In-DOTA-F3B was very limited (0.7% ID/g), while high level of radioactivity was observed in kidneys and bladder. Tumor uptake was significantly higher for ^{111}In-DOTA-F3B (2.0% IG/g) in relation to ^{111}In-DOTA-control oligonucleotide (0.7% ID/g) with tumor to muscle ratio of 4.0. These results support previous works confirming that various chelators could significantly affect the *in vivo* biodistribution of the radiolabeled compound. The tumor accumulation was confirmed by *ex vivo* scintigraphic images and autoradiography. These results pointed out that F3B aptamer is of interest for tumor imaging and for radionuclide therapy, since DOTA-F3B can be associated with a variety of beta or alpha emitters as Lutetium 177, Yttrium 90 or Bismuth 213.

The human epidermal growth factor receptor (HER) family promote tumorigenesis via cell proliferation, differentiation, migration, survival and adhesion. It includes four members: EGFR (HER1), HER2, HER3 and HER4. The aptamer U2 was selected against the epidermal growth factor receptor variant III (EGFRvIII), the most common EGFR mutant in glioblastoma. U87-EGFRvIII cells that over-express EGFRvIII were used for cell-SELEX and U87MG cells were introduced for counter selection. The aptamer U2 was directly radiolabeled with ^{188}Re (labeling efficiency of 68%) in the presence of stannous chloride. The radiolabeled aptamer significantly targeted EGFRvIII over-expressing glioblastoma xenografts in mice. ^{188}Re-U2 showed an excellent tumor uptake 3h after injection, while the radiolabeled control aptamer and free ^{188}Re accumulated mainly in the liver (Wu et al., 2014). To target brain glioblastoma the authors considered incorporating the radiolabeled aptamer in nanoparticles systems to cross the blood-brain-barrier.

The RNA aptamer E07, addressed to the extracellular domain of EGFR (Kd = 2.4 nM), was conjugated to hollow gold nanospheres (HAuNS) by annealing SH-terminated single-stranded DNA to the HAuNs first and adding the aptamer, which has 22 bases complementary to the single-stranded DNA. The aptamer-HAuNS conjugate was labeled with ^{111}In (radiochemical purity greater than 95%) using DTPA as a chelator. SPECT imaging and biodistribution in human OSC-19 oral tumors-bearing nude mice showed specific binding and higher tumor uptake of ^{111}In-DTPA-Apt-HAuNS in relation to anti-EGFR antibody conjugated nanospheres (Melacon et al., 2014).

To facilitate *in-vivo* biological applications, the RNA aptamer E07 was truncated to 48-mer and modified with 2'-fluoro at the pyrimidines to generate MinE07. An aptamer-based PET probe to evaluate EGFR expression was generated using an alkyne-modified MinE07, denoted as ME07. ^{18}F-fluorobenzoyl (FB) azide was employed as a synthon to produce ^{18}F-FB-ME07, via click chemistry. The ability of ^{18}F-FB-ME07 to image and quantify EGFR expression was tested in mice bearing A431, U87MG, and HCT-116 tumors, which presents high, medium and no expression of EGFR, respectively. In static PET imaging, despite high uptake in the liver and kidneys, ^{18}F-FB-ME07 showed accumulation in A431 tumors (1.02 %ID/g at 30 min after injection), significantly higher than the uptake in U87MG (0.53 %ID/g), and HCT-116 (0.34 %ID/g) tumors. In A431 xenografted mice, tumor/blood and tumor/muscle ratios were 3.89 and 8.65, respectively (Cheng et al., 2015).

HER2 expression is frequently increased in several types of human tumors, including those of breast cancer, head and neck, prostate, and ovary. A modified RNA aptamer with HER2-specific binding was conjugated to HYNIC and labeled with ^{99m}Tc using tricine as coligand. The complex was evaluated for diagnostic imaging of ovarian cancer cells (SKOV-3). The aptamer was modified with a 3' inverted thymidine, a primary amine on the

5' end attached through a 6-carbon alkyl linker and fluoration of 2'OH groups of ribose of pyrimidines. While the radioconjugated aptamer showed specific binding to the HER2 receptor on cells *in vitro*, no significant tumor-to-blood or tumor-to-muscle ratios were observed in tumor-bearing mice (Varmira et al., 2013). The tumor uptake was 1.9% at 1 h and 0.66% at 4 h. The authors hypothesized that this might be due to the use of tricine as the coligand and investigated the use of EDDA instead. In the animal biodistribution study, uptake of the EDDA-co-liganded ^{99m}Tc-HYNIC-RNA aptamer by the liver and spleen was lower than the aptamer with tricine. Despite no increase in tumor uptake was observed (1.4% at 1 h and 0.65% at 4 h), the tumor-to-muscle and tumor-to-bone ratios were more than three, at 1 and 4 h after injection (Varmira et al., 2014), indicating that EDDA was superior to tricine as co-ligand.

Two HER2 aptamers (HeA2_3 and HeA2_1) were conjugate to maleimide-NOTA and radiolabeled with ^{68}Ga. The diagnostic potential of these radiolabeled HER2 aptamers was evaluated by PET/MRI in mice bearing a HER2-positive and HER2-negative tumor. High uptake in blood, tissues, and organs were verified. It was explained by the slow blood clearance due to non-specific binding of the radiotracers to blood proteins. *Ex vivo* biodistribution revealed high uptake of ^{68}Ga radiolabeled HER2 aptamers in HER2 positive tumors (7.9% ID/g and 7.1% ID/g for HeA2_3 and HeA2_1, respectively) and 1.5 fold higher uptake in HER2-positive tumors versus HER2-negative tumors. However, the same behavior was observed for the negative control aptamer and no significant differences between the uptake of HER2 aptamers and a negative control aptamer were verified for both tumor types. These results suggest that tumor uptake was more related to the tumor composition factors (such as vascular network, lymphatic system, necrosis and cellular composition) than to the expression of HER2 (Gijs et al., 2016).

The SH-1194-35 is a modified DNA aptamer with a 5' amine group that contains a napthyl nucleoside to increase affinity and serum half-life. The aptamer that specifically recognizes human HER2 (Mahlknecht et al., 2013) was radiolabeled with ^{18}F-fluoride using (^{18}F-SFB) and tested in BT474 (HER2-positive human breast cancer cells) tumor-bearing mice. The biodistribution showed a radiolabeled aptamer tumor uptake of 0.62% ID/g at 1 h post injection. The two major excretory systems are kidneys and intestine. PET images taken 120 min post injection showed the tumor clearly labeled. BT474 tumors (HER2-positive) showed significantly higher uptake than MB-MDA231 tumors (HER2-negative). The ^{18}F-labeled aptamer enabled appropriate visualization of HER2 expression by human breast cancer cells (Kin et al., 2019).

Protein tyrosine kinase-7 (PTK7) was initially identified as colon carcinoma kinase-4 (CCk4). PTK7 is a member of receptor tyrosine kinase superfamily that is highly expressed in various human malignancies including colon and gastric cancers, myeloid leukemia, lung cancer and glioblastoma overexpressing CD44. The DNA aptamer Sgc8, generated by cell-SELEX on human lymphoblastic leukemia cells was identified to interact with PTK7 (Shi et al., 2011). This aptamer was evaluated as a PET radiotracer in mice bearing subcutaneous tumors from HCT116 (which express high levels of PTK7) and U87MG (that express lower levels of PTK7) cell lines. Radiolabeling was performed using an alkynyl functionalized Sgc8 aptamer and ^{18}F-fluorobenzyl via click chemistry. PET studies showed specific accumulation of the ^{18}F-Sgc8 in HCT116 tumors (0.76% ID/g at 30 min) and lower uptake by U87MG tumors (0.13% ID/g). ^{18}F-Sgc8 uptake was also higher than that of the control sequence in both models. In addition, the uptake by liver metastases from HCT116 cells was higher than by the subcutaneous tumor. The radiolabeled aptamer was rapidly cleared from the blood through the renal route to give high tumor-to-blood and tumor-to-muscle ratios of 7.29 and 10.25, respectively (Jacobson et al., 2015).

Wang et al., (2015) labeled the Sgc8 aptamer with ^{18}F-arenes by click chemistry using an oxygen ortho-stabilized iodonium derivative (OID). *In vivo* studies were performed by administration of radiolabeled aptamer to female nude mice with HCT116 tumors followed by PET imaging. Clear tumor visualization was possible at all-time points, with tumor uptake of 0.71% ID/g and tumor-to-muscle ratios of 3.54 and 4.15 at 1 h and 2 h post injection. The relatively modest tumor uptake of aptamers was related to their rapid clearance via urinary elimination.

The aptamer Sgc8-c is a truncated sequence (41-mer) of the original DNA aptamer Sgc8, which showed similar binding properties. Sgc8-c DNA was tagged with the fluorophore AlexaFluor647 or radiolabeled with ^{99m}Tc using the chelating moiety HYNIC. For this purpose, the 5' end of the aptamer was modified with an amine-terminated 6-carbon spacer. *In vivo* studies in melanoma (B16F1 cells) and lymphoma (A20 cells) animal tumor models were performed. B16F1-bearing mice injected with Sgc8-c-Alexa647 showed little tumor uptake, while A20-bearing mice present high tumor uptake. No tumor uptake was observed after injection of Sgc8-c-^{99m}Tc in either mouse model (Calzada et al., 2017). In addition to a suitable molecular weight for renal filtration (40-60 kDa), Sgc8-c-^{99m}Tc probe is more hydrophilic than the fluorescent counterpart Sgc8-c-Alexa647, which led to greater clearance. This result highlighted that small physicochemical differences in probe composition could yield large differences *in vivo* behavior.

A recent area of investigation is the use of radiolabeled aptamers probes for bacterial infection detection. If detected early, most infections can be cured with proper treatment, but a delayed diagnosis is associated with higher mortality. The sensitivity of nuclear medicine imaging techniques makes it a suitable tool for the specific diagnosis of focal infections. SPECT and PET platforms allow whole-body imaging, which is important when there are no localizing signs. A variety of radiopharmaceuticals has been used to detect infection, but long-term clinical use has shown that most of them cannot distinguish between aseptic inflammation and infection (Ferro e Flores et al., 2012). The main challenges regarding scintigraphic infection imaging are to distinguish between infection and sterile inflammation and selectively distinguish among different types of infection for choosing the correct therapy.

Two studies used the *Staphylococcus aureus*-specific aptamers SA20, SA23 and SA34 (Cao et al., 2009) for infection foci identification. The radiotracer was constituted of a mixture of the three aptamers, since the combined use of these aptamers has an additive effect on the *S. aureus* recognition compared to using only a single aptamer. The labeling with ^{99m}Tc was performed by the direct method (Correa et al., 2014). Santos et al., (2015) performed an *ex vivo* biodistribution study in which the radiolabeled aptamers were able to identify the *S. aureus* infection foci displaying a target/non-target ratio of 4.0. For the control mice group infected with *Candida albicans* this ratio was 2.0, while for mice with aseptic inflammation it was 1.2. In the work of Santos et al., (2017), scintigraphic images of *S. aureus* infectious foci were obtained using the same radiotracer. The target/non-target ratio determined by ROI analysis of images were 4.6 and 4.5 at 1 h and 3 h, respectively. For the control group consisting of *S. aureus*-infected animals that were inoculated with a ^{99m}Tc-radiolabeled library (oligonucleotides with random sequences) these ratios were of 1.6 at both times. The image profiles showed that the urinary tract was the main route of radiotracer elimination.

Ferreira et al., (2014) selected the aptamer Antibac1 for the peptidoglycan, the main component of the bacterial cell wall. The aptamer showed binding capacity for *S. aureus* and *E. coli* cells *in vitro,* but the binding to *C. albicans* and human fibroblasts was negligible. Graziani et al., (2017) demonstrated that Antibac1 binds with high efficiency to Gram-positive and Gram-negative bacterial species and displayed a Kd for *S. aureus* cells of 170 nM. This aptamer can be considered as a generic probe for bacteria identification. The Antibac1 was directly labeled with ^{99m}Tc and evaluated for bacterial infection diagnosis by scintigraphy. The biodistribution and scintigraphic imaging studies with the ^{99m}Tc-Antibac1 were carried out in two different experimental infection models: Bacterial-infected mice (*S. aureus)* and fungal-infected mice (*C. albicans*). A ^{99m}Tc-radiolabeled library was used as a control for both models. Scintigraphic images of *S. aureus*-infected mice at 1.5 and 3.0 h after ^{99m}Tc-Antibac1 injection showed target to non-target ratios of 4.7 and 4.6, respectively. These values were statistically higher than those achieved for the ^{99m}Tc-library at the same periods (1.6 and 1.7, respectively). Noteworthy, ^{99m}Tc-Antibac1 and ^{99m}Tc-library showed similar low target to non-target ratios in the fungal-infected model: 2.0 and 2.0 for ^{99m}Tc-Antibac1 and 2.1 and 1.9 for ^{99m}Tc-library, at the same times. These findings suggest that the ^{99m}Tc-Antibac1 is a feasible imaging probe to identify a bacterial infection focus. In addition, this radiolabeled aptamer seems to be suitable in distinguishing between bacterial and fungal infection (Ferreira et al., 2017). The findings suggested that aptamers could be more extensively studied to develop specific infection diagnostic radiopharmaceuticals for different categories of pathogens.

Conclusion

Nowadays, there is a great interest in the development of more specific radiopharmaceuticals, able to identifiy a specific type of cancer cell, specific microorganism or a molecular marker of a particular disease. Because of that, aptamers are rising as a trend field in nuclear medicine. Aptamer-based radiopharmaceuticals may also contribute towards personalized medicine allowing monitoring the efficacy of treatments, by developing an unique aptamer for a specific patient. Given that, currently, 95% of radiopharmaceuticals are used for diagnostic purposes, much still can be explored in terms of therapy or theranostics applications by using aptamers radiolabeled with the appropriate radionuclides.

The first application of aptamers as radiopharmaceutical was published in 1997 using ^{99m}Tc to radiolabel an aptamer specific to neutrophil elastase. Since that, the number of publications and study areas has increased considerably. The progress made on aptamer technology in the field of nuclear medicine in the last decade indicates that these molecules will soon proceed to clinical applications, leading to an aptamer-based radiopharmaceutical approved for use in a near future.

References

Boisgard, R., Jego, B., Siquier, K., Hinnen, F., Dollé F., Friebe, M., Borkowski S., Dinkelborg L., Tavitian, B. (2009). *In vivo* PET tumour imaging using an [F-18] labelled aptamer targeting tenascin-C. *J Nucl Med*, 50, 1594.

Borbas, K. E., Ferreira, C. S. M., Perkins, A., Bruce, J. I., Missailidis, S. (2007). Design and synthesis of mono- and multimeric targeted radiopharmaceuticals based on novel cyclen ligands coupled to anti-MUC1 aptam ers for the diagnostic imaging and targeted radiotherapy of cancer. *Bioconjugate Chem*, 18, 1205-1212.

Bouvier-Müller, A. and Ducongé, F. (2018). Application of aptamers for *in vivo* molecular imaging and theranostics. *Adv Drug Deliv Rev*, 134, 94-106.

Cai, S., Yan, J., Xiong, H., Liu, Y., Peng, D., Liu, Z. (2018). Investigations on the interface of nucleic acid aptamers and binding targets. *Analyst,* 143, 5317-5338.

Calzada, V., Moreno, M., Newton, J., González, J., Fernández, M., Gambini, J. P., Ibarra, M., Chabalgoity, A., Deutscher, S., Quinn, T., Cabral, P., Cerecetto, H. (2017). Development of new PTK7-targeting aptamer-fluorescent and radiolabelled probes for evaluation as molecular agents: Lymphoma and melanoma *in vivo* proof of concept. *Bioorg Med Chem,* 25, 1163-1171.

Cao, X., Li, S., Chen, L., Ding, H., Xu, H., Huang, Y., L,i J., Liu, N., Cao, W., Zhu, Y., Shen, B., Shao, N. (2009). Combining use of a panel of ssDNA aptamers in the detection of *Staphylococcus aureus*. *Nucleic Acid Res*, 37, 4621-4628.

Carmo, F. S., Ricci-junior, E., Cerqueira-Coutinho, C., Albernaz, A. S., Bernardes, E. S., Missailidis, S., Oliveira-Santos, R. (2017). Anti-MUC1 nano-aptamers for triple-negative breast cancer imaging by single-photon emission computed tomography in inducted animals: initial considerations. *Int J Nanomedicine*, 12, 53-60.

Carter, L. M., Poty, S., Sharma, S. K., Lewis, J. S. (2017). Preclinical optimization of antibody-based radiopharmaceuticals for cancer imaging and radionuclide therapy-Model, vector, and radionuclide selection. *J Labelled Comp Radiopharm*, 61, 611-635.

Charlton, J., Sennello, J., Smith, D. (1997). In vivo imaging of inflammation using an aptamer inhibitor of human neutrophil elastase. *Chem Biol*, 4, 809-816.

Cheng, S., Jacobson, O., Zhu, G., Chen, Z., Liang, S. H., Tian, R., Yang, Z., Niu, G., Zhu, X., Chen, X. (20190. PET imaging of EGFR expression using na ^{18}F-labeled RNA aptamer. *Eur J Nucl Med Mol Imaging*, 46, 948-956.

Correa, C. R., de Barros A. L. B., Ferreira, C. A., Goes, A. M., Cardoso, V. N., Andrade, A. S. R. (2014). Aptamers directly radiolabeled with technetium-99m as a potential agent capable of identifying carcinoembryonic antigen (CEA) in tumor cells T84, *Bioorg Med Chem Lett*, 24, 1998-2001.

Cuocolo, A., Pappatà, S., Zampella, E., Assante, R. (2018). Advances in SPECT Methodology. *Int Rev Neurobiol,* 141, 77-96.

Daniels, D. A., Chen, H., Hicke, B. J., Swiderek, K. W., Gold, L. (2003). A tenascin-C aptamer identified by tumor cell SELEX: Systematic evolution of ligands by exponential enrichment. *Proc Natl Acad Sci USA.,* 100, 15416-15421.

Da Rocha Gomes, S., Miguel, J., Azema, L., Eimer, S., Ries, C., Dausse, E., Loiseau, H., Allard, M. (2012). ^{99m}Tc-MAG3-aptamer for imaging human tumor associated with high level of matrix metalloprotease-9. *Bioconjug Chem,* 23, 2192-200.

Davydova, A,, Vorobjeva, M., Pyshnyi, D., Altman, S., Vlassov, V., Venyaminova, A. (2016). Aptamers against pathogenic microorganisms. *Crit Rev Microbiol*, 42, 847-65.

de Almeida, C. E. B., Alves, L. N., Paulino, E. T., Cabral-Neto, P. J., Missailidis, S.(2017). Aptamer delivery of siRNA, radiopharmaceutics and chemotherapy agents in cancer. *Int J Pharm*, 525, 334-342.

de Blois, E., Chan, H. S., Naidoo, C., Prince, D., Krenning, E. P., Breeman, W. A. P. (2011). Characteristics of SnO2-based 68Ge/68Ga generator and aspects of radiolabelling DOTA-peptides. *Appl Rad Isot*, 69,308–315.

de Vries, E. F., Vroegh, J. (2003). Evaluation of fluorine-18-labeled alkylating agents as potential synthons for the labeling of nucleotides. *Appl Radiat Isot*, 58, 469-476.

Dougan, H., Weitz, J. I., Stalfford, A. R., Gillespie, K. D., Klement, P., Hobbs, J. B., Lyster, D. M. (2003). Evaluation of DNA aptamers directed to thrombin as potential thrombus imaging agents. *Nucl Med Biol,* 30, 61-72.

Famulok, M. (1994). Molecular recognition of amino acids by RNA aptamers: An l-citruline binding RNA motif and its evolution into a l-arginine binder. *J Am Chem Soc*, 116, 1698-1706.

Farzin, L., Shamsipur, M., Moassesi, M. E., Sheibani, S. (2019). Clinical aspects of radiolabeled aptamers in diagnostic nuclear medicine: A new class of targeted radiopharmaceuticals. *Bioorganic & Medicinal Chemistry,* 27, 2282-2291.

Ferro-Flores, G., Ocampo-Garcia, B. E., Melendez-Alafort, L. (2012) Development of specific radiopharmaceuticals for infection imaging by targeting infectious micro-organisms. *Curr Pharm Des*, 18, 1098-1106.

Ferreira, C. D., Matthews, C. S., Missailidis, S. (2006). DNA aptamers that bind to MUC1 tumor marker: design and characterization of MUC1-binding single-stranded DNA aptamer. *Tumor Biol*, 27, 289-301.

Ferreira, I. M., de Sousa Lacerda, C. M., Faria, L. S., Corrêa, C. R., Andrade, A. S. R. (2014). Selection of Peptidoglycan-Specific Aptamers for Bacterial Cells Identification, *Appl Biochem Biotechnol*, 174, 2548–2556.

Ferreira, I. M., de Sousa Lacerda, C. M., dos Santos, S. R., de Barros, A. L. B, Fernandes, S. O., Cardoso, V. N., Andrade, A. S. R. (2017). Detection of bacterial infection by a technetium-99m-labeled peptidoglycan aptamer. *Biomed Pharmacother*, 93, 931-938.

Gao, H., Qjan, J., Cao, S., Yang, Z., Pang, Z., Pan, S., Fan, L., Xi, Z., Jiang, X., Zhang, Q. (2012). Precise glioma targeting of and penetration aptamer and peptide dual-functioned nanoparticles. *Biomaterials,* 33, 5115-23.

Geiger, A., Burgstaller, P., von der Eltz, H., Roeder, A., Famulok, M. (1996). RNA aptamers that bind L-arginine with sub-micromolar dissociation contants and high enantioselectivity. *Nucleic Acids Res,* 24, 1029-36.

Gijs, M., Aerts, A., Impens, N., Baatout, S., Luxen, A. (2016). Aptamers as radiopharmaceuticals for nuclear imaging and therapy. *Nucl Med Biol,* 43, 253-271.

Gijs, M., Becker, G., Plenevaux, A., Bahri, M. A., Aerts. An M., Impens, N., Baatout, S., Luxen, A. (2016). Biodistribution of novel 68Ga-radiolabelled HER2 aptamer in mice. *J Nucl Med Radiat Ther*, 7, 1000300.

Girvan, A. C., Teng, Y., Casson, L. K. Thomas, S. D., Jüliger, S, Ball, M. W., Klein, J., B., Pierce, W. M. Jr., Barve, S. S., Bates, P. J. (2006). AGRO100 inhibits activation of nuclear factor-κB (NF-κB) by forming a complex with NF-κB essential modulador (MEMO) and nucleolin. *Mol Cancer Ther*, 5, 1790-1799.

Graziani, A. C., Stets, M. I., Lopes, A. L.,. Schluga, P. H., Marton, S., Mendes, I. F., Andrade, A. S. R., Krieger, M. A., Cardoso, J. (2017). High efficiency binding aptamers for a wide range of sepsis bacterial agents. *J Microbiol Biotechnol,* 27, 838–843.

Hassanzadeh, L., Chen, S., Veedu, R. N. (2018). Radiolabeling of nucleic acid aptamers for highly sensitive disease-specific molecular imaging. *Pharmaceuticals,* 11, 106.

Healy, J., Lewis, S., Kurz, M., Boomer, R., Thompson, K., Wilson, C., McCauley, T. G. (2004). Pharmacokinetics and biodistribution of a novel aptamer compositions. *Pharm Res*, 21, 2234-46.

Henry, S. P, Johnson, M, Zanardi, T. A., Fey, R., Auyeung, D., Lappin, P. B., Levin, A. A. (2012). Renal uptake and tolerability of a 2′-O-methoxethyl modified antisense oligonucleotide (ISIS 113715) in monkey. *Toxicology,* 301, 13-20.

Hiecke, J. H., Marion, C., Chang, Y. F., Gould, T., Lynott, C. K., Parma, D., Schimidt, P. G., Warrens, S. (2001). Tenascin-C aptamers are generated using tumor cells and purified protein. *J. Biol Chem*, 276, 48644-48654.

Hicke, B. J., Stephens, A. W., Gould, T., Chang, Y. F., Lynott, C. K., Heil, J., Borkowski, S., Hilger, C. S., Cook, G., Warren, S., Schmidt, P. G. (2006). Tumor targeting by an aptamer. *J Nucl Med*, 47, 668-78.

Hong, H., Goel, S., Zhang, Y., Cai, W. (2011). Molecular Imaging with Nucleic Acid Aptamers. *Curr Med Chem*, 18, 4195-4205.

Hwang, D. W., Ko, H., Y., Lee, J. H., Kang, H., Ryu, S. H., Song, I. C., Lee, D. S., Kim, S. (2010). A nucleolin-targeted multimodal nanoparticle-imaging probe for tracking cancer cells using an aptamer. *J. Nucl Med,* 51, 98-105.

Jacobson, O., Yan, X., Niu, G., Weiss, I. D., Ma, Y., Szajek, L. P., Shen, B., Kiesewetter, D. O., Chen, X. (2015). PET imaging of tenascin-C with a radiolabeled single-stranded DNA aptamer. *J Nucl Med*, 56, 616-621.

Jacobson, O., Weiss, I. D., Wang, L., Wang, Z., Yang, X., Dewhurst, Y. M., Zhu, G., Niu, G., Kiesewetter, D. O., Vasdev, N., Liang, S. H., Chen, X. (2015). ^{18}F-labeled single-stranded DNA aptamer for PET imaging of protein tyrosine kinase-7 expression. *J Nucl Med*, 56, 1780-1785.

Jenilson, R. D., Gill, S. C., Pardi, A., Polisky, B. (1994). High-resolution molecular discrimination by RNA. *Science*, 263, 1425-9.

Kang, W. J., Lee, J., Lee, Y. S., Cho, S., Ali, B, A., Al-Khedhairy, A. A., Heo, H., Kim, S. (2015) Multimodal imaging probe for targeting cancer cells using uMUC-1 aptamer. *Colloids Surf B Bionterfaces*, 136, 134-140.

Khan, H., Missailidis, S. (2008). Aptamers in oncology: A diagnostic perspective. *Gene Ther \mol Biol,* 12, 111-128.

Kin, H. J., Park, J. Y., Lee, T. S., Song, I.,H., Cho, Y. L., Chae, J. R., Kang, H. Lim, J. H., Lee, J. H., Kang, W. J. (2019). PET imaging of HER2 expression with as ^{18}F-fluoride labeled aptamer. *Plos One*, 14, e0211047.

Kryza, D., Debordeasux, F., Azéma, L., Hassan, A., Paurelle, O., Schulz, J., Savona-Baron, C., Charignon, E., Bonazza, P., Taleb, J., Fernandez, P., Toulmé, J. J. (2016). *Ex vivo* and *in vivo* imaging and biodistribution os aptamers targeting the human matrix metalloprotease-9 in melanomas. *Plos One*, 11, e0149387.

Li, J., Zheng, H., Bates, P. J., Malik, T., Li, X. F., Trent, J. O. Chin, K. N. (2014). Aptamer imaging with Cu-64 labeled AS1411: Preliminary assessment in lung cancer. *Nucl Med Biol*, 41, 179-185.

Lipi, F., Chen, S. (2016). In vitro evolution of chemically-modified nucleic acid aptamers: Pros and cons, and comprehensive selection strategies. *RNA boil*, 13, 1232-1245.

Liu, Z., Li, Z. B., Cao, Q., Liu, S., Wang, F., Chen, X. (2009). Small-animal PET of tumors with (64)Cu-labeled RGD-bombesin heterodimer. *J Nucl Med,* 50, 1168-77.

Liu, Y., Liu, G., Hnatowich, D. J. (2010). A Brief Review of Chelators for Radiolabeling Oligomers. *Materials* 3, 3204-3217.

Maecke, H. R., Hofmann, M., Haberkorn, U. (2005). 68Ga-Labeled Peptides in Tumor Imaging. *J Nucl Med*, 46, 172–178.

Mahlknecht, G., Maron, R., Mancini, M., Schechter, B., Sela, M., Yaden, Y. (2013). Aptamer to ErbB-2/HER2 enhances degradation of the target and inhibits tumorigenic growth. *Proc Natl Acad Sci USA*, 110, 8170-5.

Mariani, G., Bruselli, L., Kuwert, T., Kim, E. E., Flotats, A., et al., (2010). A review on the clinical uses of SPECT/CT. *Eur J Nucl Med Mol Imaging*, 37, 1959–1985.

Melacon, M. P., Zhou, M., Zhang, R., Xiong, C., Allen, P., Wen, X., Huang, Q., Wallace, M., Myers, J, N., Stafford, R. J., Liang, D., Ellington, A., D., Li, C. (2014). Selective uptake and imaging of aptamer- and antibody-conjugated hollow nanospheres targeted to epidermal growth factor receptors overexpressed in head and neck cancer. *ACS Nano*, 8, 4530-8.

Missailidis, S., Perkins, A. (2007). Aptamers as novel radiopharmaceuticals: Their applications and future prospects in diagnosis and therapy. *Cancer Biother and Radiopharm*, 22, 453-468.

Mosing, R. K., Bowser, M. T. (2009). Isolating aptamers using capillary electrophoresis-SELEX (CE-SELEX). *Methods Mol Biol*, 535, 33-43.

Muller-Bouvier, A., Ducongé, F. (2018) Application of aptamers for in vivo Molecular Imaging and Theranostics. *Adv. Drug Deliv. Rev, 134, 94-106*

Noaparast, Z., Hosseinimehr, S. J., Piramoon, M., Abedi, S. M. (2015). Tumor targeting with a ^{99m}Tc-labeled AS1411 aptamer in prostate tumor cells. *J Drug Target*, 23,497-505.

Nutt, R. (2002). The History of Positron Emission Tomography. *Mol Imag Bio*, 4, 11-26.

Oliveira, R., Santos, D., Ferreira, D., Coelho, P., Veiga, F. Preparações radiofarmacêuticas e suas aplicações. *Braz J Pharm Sci*, 42, 151-165.

Orit, J., Xuefeng, Y. (2015). PET imaging of tenascin-c with a radiolabeled single-stranded DNA aptamer. *J Nucl Med*, 56,616-621.

Park, J. Y & Lee, T. S. (2016) Hybridization-based aptamer labeling using complementary oligonucleotide platform for PET and optical imaging. *Biomaterials*, 100, 143-151.

Pascuak, L., Cerqueira-Coutinho, C., Garcia-Fernández, A., Luis, B., Bernardes, E. S., Albernaz, M. S., Missailidis, S., Martínez-Máñez, R., Santos-Oliveira, R., Orzaez, M., Sancenón, F. (2017). MUC1 aptamer-capped mesoporous silsica nanoparticles for controlled drug delivery and radio-imaging applications. 13, 2495-2505.

Pendergrast, P., Marsh, H., Grate, D., Healy, J., Stanton, M. (2005). Nucleic acid aptamers for target validation and therapeutic applications. *J Biomol Technol*, 16, 224-34.

Phelps, M. (2000). Positron emission tomography provides molecular imaging of biological processes. *PNAS,* 97, 9226-9233.

Pieve, C. D., Perkins, A. C., Missailidis, S. (2009). Anti-MUC1 aptamers: radiolabeling with (99mTc) and biodistribution in MCF-7 tumor bearıng mice. *Nucl Med Biol*, 36,703-10.

Pieve, C. D., Blackshaw, E., Missaililidis, S., Perkins, A. C. (2012). PEGylation and biodistribution of an anti-MUC1 aptamer in MCF-7 tumor bearing mice. *Bioconjug Chem*, 23, 1377-81.

Powsner, R. A., Powsner, E. R. *Essential Nuclear Medicine Physics*. Second edition, 2006. Blackwell.

Pretze, M., Pietzsch, D., Mamat, C. (2013). Recent Trends in Bioorthogonal Click-Radiolabeling Reactions Using Fluorine-18. *Molecules,* 18, 8618-8665.

Price, E. W., Orvig, C. (2014). Matching chelators to radiometals for radiopharmaceuticals. *Chem Soc Rev*, 43, 260-290.

Rosenberg, J. E., Bambury, R. M., Van Allen, E. M., Drabkin, H. A., Lara, P. N., Harzstark, A. L., Wagle, N., Figlin, R. A., Smith, G. W., Garraway, L. A., Choueiri, T., Erlandsson, F., Laber, D. A. (2014). A phase II trial of AS1411 (a novel nucleolin-targeted DNA aptamer) in metastatic renal cell carcinoma. *Invest New Drug*, 32, 179-87.

Rotstein, B. H., Stepherson, N. A., Vasdev, N., Liang, S. H. (2014). Spirocyclic hypervalent iodine (III)-mediated radiofluorination of non-activated and hindered aromatics. *Nat Commun*, 5, 4365.

Sa, L. T., Simmons, S., Missailidis, S., da Silva, M. I., Santos-Oliveira, R. (2013). Aptamer-based nanoparticles for cancer targeting. *J Drug Target*, 21,427-34.

Saha, G. B. *Fundamentals of Nuclear Pharmacy*. Sixth edition, 2010. Springer.

Saha, G. B. *Physics and Radiobiology of Nuclear Medicine*. Third edition, 2006. Springer.

Santos, S. R., Corrêa, C. R., De Barros A. L. B., Serakides, R., Fernandes, S. O., Cardoso, V. N., Andrade, A. S. R. (2015) Identification of *Staphylococcus aureus* infection by aptamers directly radiolabeled with technetium-99 m. *Nucl Med Biol*, 42, 292–298.

Santos, S. R., De Sousa Lacerda, C. M., Ferreira, I. M., Barros, A. L. B., Fernandes, S. O., Cardoso, V. N., Andrade, A. S. R. (2017). Scintigraphic imaging of Staphylococcus aureus infection using ^{99m}Tc radiolabeled aptamers. *Appl Radiat Isto*, 128, 22-27.

Sassanfar, M., Szostak, J. W. (1993). An RNA motif that binds ATP. *Nature,* 364, 550-3.

Sefah, K., Shangguan, D., Xiong, X., O'Donoghue, M. B., Tan, W. (2010). Development of DNA aptamers using Cell-SELEX. *Nat. Protoc*, 5, 1169–1185.

Seo, J. W., Zhang, H., Kukis, D. L., Meares, C. F., Ferrara, K. W. (2008). A Novel Method to Label Preformed Liposomes with 64Cu for Positron Emission Tomography (PET) Imaging. *Bioconjugate Chem*, 19, 2577–2584.

Sharp, P. F., Gemmel, H. G., Murray, A. D. *Practical Nuclear Medicine.* Third edition, 2005. Springer.

Shi, H., He, X., Wang, K., Wu, X., Ye, X., Guo, Q., Tan, W., Qing, Z., Yang, X., Zhou, B. (2011). Activable aptamer probe for contrast-enhanced in vivo cancer imaging based on cell membrane protein-triggered conformation alteration. *Proc Natl Acad Sci USA,* 108, 3900-3905.

Signore, A., Annovazzi, A., Chianelli, M., Corsetti, F., Van de Wiele, C., Watherhouse, R. N. (2001). Peptide radiopharmaceuticals for diagnosis and therapy. *Eur J Nucl Med.* 28, 1555-1565.

Tan, S. Y., Acquah, C., Sidhu, A., Ongkudon, C. M., Yon, L. S., Danquah, M. K. (2016). SELEX Modifications and Bioanalytical Techniques for Aptamer-Target Binding Characterization. *Crit Rev Anal Chem,* 46(6), 521-37.

Tavitian, B. (2009). *In vivo* PET tumour imaging using an [F-18] labelled aptamer targeting tenascin-C. *J Nucl Med,* 50, 1594.

Thiviyanathan, V. and Gorenstein, D. G. (2012). Aptamers and the Next Generation of Diagnostic Reagents. *Proteomics Clin App.*, 6, 563–573.

Ting, G., Chang, C. H., Wang, H. E. (2009). Cancer Nanotargeted Radiopharmaceuticals for Tumor Imaging and Therapy. *Anticancer Res,* 29, 4107-4118.

Tuerk, C. Gold, L. (1990). Systematic evolution of ligands by exponential enrichment – RNA ligands to Bacteriophage-T4. *Science* 249 (4968), 505-510.

Varmira, K., Hosseinimehr, S. J., Noaparast, Z., Abedi, S. M. (2013). A HER2-targeted RNA aptamer molecule labeled with ^{99m}Tc for single-photon imaging in malignant tumors. *Nucl Med Biol,* 40, 980-986.

Varmira, K., Hosseinimehr, S. J., Noaparast, Z., Abedi, S. M (2014). An improved radiolabelled RNA aptamer molecule for HER2 imaging in cancers. *J Drug Target*, 22, 116-22.

Wang, L., Jacobson, O., Avdic, D., Rotstein, B. H., Weiss, I. D., Collier, L., Xiaoyuan, C., Vasdev, N., Liang, S. H. (2015). *Ortho*-stabilized ^{18}F-azido click agents and application in PET imaging of single-stranded DNA aptamer. *Angew Chem Int Ed England*, 54, 12777-12781.

Wang, T., Chen, C., Larcher, L. M., Barrero, R. A. and Veedu, R. N. (2019). Three decades of nucleic acid aptamer technologies: Lessons learned, progress and opportunities on aptamer development. *Biotechnology Advances*, 37(1), 28-50.

Wei, L,, Butcher, C., Miao, Y., Gallazzi, F., Quinn, T. P., Welch, M. J., Lewis, J. S. (2007). Synthesis and biologic evaluation of 64Cu-labeled rhenium-cyclized alpha-MSH peptide analog using a cross-bridged cyclam chelator. *J Nucl Med*, 48, 64-72.

White, R. R., Sullenger, C. P., Rusconi, C. P. (2000). Developing aptamers into therapeutics. *J Clin Invest,* 106, 929-934.

Willian, M., Rockey, B. (2011). Synthesis and radiolabeling of chelator-RNA aptamer bioconjugates with copper-64 for targeted molecular imaging. *Bioorg Med Chem*, 19, 4080-4090.

Wu, X., Liang, H., Tan, Y., Yuan, C., Li, S., Li, X., Li, G., Shi, Y., Zhang, X. Cell-SELEX aptamer for highly specific radionuclide molecular imaging of glioblastoma *in vivo*. *PLoS One*, 9, e90752.

Yousefi, M., Dehghani, S., Nosrati, R., Zare, H., Evazalipour, M., Mosafer, J., Tehrani, B. S., Pasdar, A., Mokhtarzadeh, A., Ramezani, M. (2019). Aptasensors as a new sensing technology developed for the detection of MUC1 mucin: A review. *Biosens Bioelectron*, 130, 1-19.

Zolle, I. *Technetium-99m Pharmaceuticals: preparation and quality control in nuclear medicine*. 2007. Springer.

In: A Comprehensive Guide to Aptamers ISBN: 978-1-53616-293-6
Editor: Tom Shuster

Chapter 3

APTAMERS VS. ANTIBODIES IN DIFFERENT DETECTION TECHNIQUES

Alena K. Ryabko**, Maksim A. Marin, Natalia A. Zeninskaya, Victoria V. Firstova, PhD, and Igor G. Shemyakin**, PhD
FBIS State Research Center for Applied Microbiology, Obolensk, Russia

ABSTRACT

From the time the first aptamer was discovered with the use of SELEX technology until now the interest of the research community to this topic is growing. The possibility of obtaining oligonucleotide sequences with high affinity to a given target that recognize molecules of various nature and perform it with incredible specificity (including the possibility of obtaining aptamers able to identify small molecules that differ in one functional group), the accessibility of their synthesis and simplicity of making a wide range of modifications give a chance for using aptamers in many applications, even in those still dominated by antibodies. Thus, aptamers are now widely used in various applications: in diagnostic

* Corresponding Author's E-mail: ryabko_alena@mail.ru; igshemyakin@mail.ru.

systems (aptaPCR, PCR, biosensors, Lab On Chip, etc.), in therapeutic drugs (Macugen and a lot of drugs at the clinical trials), in research work and routine biotechnology procedures such as affinity chromatographic purification, cytometric analysis, intracellular fluorescence imaging,etc. However, despite the success of many research groups in obtaining RNA and DNA aptamers, they are still not widely used. The practice of selection and modification of aptamers has not yet replaced the adopted practice of obtaining monoclonal antibodies. In this chapter we will discuss the perspectives of using aptamers in various detection methods, their advantages and disadvantages, and the results of such work carried out until the present day.

Keywords: antibody, aptamer, detection, flow cytometry, aptasensor, aptaPCR, iaPCR, lateral flow aptamer assay

INTRODUCTION

Antibodies are widely used in theoretical and applied science. The development of technology of obtaining monoclonal antibodies with required specificity [1] has become one of turning points in the progress of biotechnology and has given unique opportunities for the use of antibodies in molecular biology, biochemistry and medicine. Nowadays it is impossible for a modern diagnostic or research laboratory not to have such a highly accurate tool as monoclonal antibodies. They are considered the "gold standard" in diagnostics and therapy. Antibodies of various types are commercially available with different modifications and as part of a huge number of therapeutic and diagnostic products.

Fifteen years after hybridoma technology was discovered and the first monoclonal antibody was produced, the technology of selecting highly specific ligands of a completely different nature - nucleotide - was announced. And such ligands were named aptamers [2]. Aptamers have a number of advantages over antibodies. For example, aptamers can be selected against practically any molecules including small ones, against non-immunogenic and even toxic substances. In addition, it is possible to obtain aptamers that distinguish low-molecular substances different in one functional group, that interact with the original substance only and not with its metabolites, and that specifically interact with one of the isomers or enantiomers in the substance [3, 4]. As aptamers are much smaller than antibodies and therefore have low immunogenicity and good bioavailability, it allows for their safer use as a drug. On the other hand, any aptamer is amenable to post-selection modification. The range of modifications of nucleotide sequences is almost limitless, and is defined only by the aim of the researcher. We can increase its molecular weight by incorporating PEG molecules to the skeleton of aptamer in order to optimize the level of blood clearance of the substance. We can also add the modified nucleotides into the sequence, or cap the 3'-end, or add non-nucleotide linkers in order to increase the stability of the aptamer and resistance of the molecule to endonuclease degradation [5, 6]. To protect aptamers from nucleases and increase their stability under various conditions and their affinity, modified nucleotides such as 2'-Fluoro-RNA, 2'-O-Methyl-RNA, 2'-NH2-RNA, L-RNA and L-DNA, Bridged Nucleic Acid (BNA) or Peptide Nucleic Acids (PNA) can be introduced into the chain. Knowing the primary sequence of the aptamer and its structure, we can modify the given sequence in order to increase its affinity and/or optimize the size of the molecule and not to lose the binding effectiveness of the aptamers when only some of the structures of obtained aptamer are responsible for the specific interaction and the rest of the sequence remain ballast. It is also possible to obtain a bivalent aptamer by combining two specific sequences [7, 8]. Changes in aptamers can be applied precisely and directly with the use of chemical synthesis. For example, for oriented immobilization or conjugation of the molecule we can incorporate the required functional groups in the aptamer. It is crucially important that the selection process is faster, simpler and cheaper than obtaining of monoclonal antibodies and is carried out exclusively in vitro. To reproduce the obtained sequence with any required number of changes and modifications is not costly either. In addition, aptamers are much more stable in a wide range of temperatures than protein molecules, and are also able to tolerate many denaturation-renaturation cycles without loss of their properties. Another advantage of aptamers over antibodies is their ability to be used as an amplifiable detecting agent. This can significantly increase the sensitivity of analyte detection in the sample. We should also take into consideration that aptamers cost much less than antibodies at all stages of production.

At the moment aptamer selection technologies are developing rapidly, many research groups offer their own simple and convenient methods. The technology becomes routine and an increasing number of researchers adopt it as a standard practice. Nowadays one can purchase such commercial kits for aptamer selection as X-Aptamer Selection Kit from AMBiotech (http://am-biotech.com/products-2/x-aptamer-selection-kit-for-molecular-targets), Aptamer Selection Starter Kit with N40 Library from InnovoGENE Biosciences (https://www.innovogene.com/store/pc/viewPrd.asp?idproduct=22), XELEX DNA and RNA Core Kits from Roboklon (https://www.roboklon.com/index.php?show=46). Commercial companies also offer nucleotides for routine modification of aptamers - Nucleotides for SELEX/Aptamer Modification by Jena Bioscience (https://www.jenabioscience.com/ nucleotides-nucleosides/ nucleotides-by-application/in-drug-discovery/selex-aptamer-modification). And it is not necessary to carry out the whole selection process by yourself - there is database of ready-made aptamer sequences such as AptaGen (https://www.aptagen.com) with the information about 540 sequences of different specificity. One can order from affiliated companies the synthesis of these sequences with the label required by the researcher. Naturally, the size of such database is significantly smaller compared to the range of commercially available antibodies, smaller by an order of magnitude. Aptamer Sciences Inc. also offers ready-made and validated aptamers as independent reagents or as part of kits for isolation of cells and targets (www.aptsci.com). The offer of this company is limited though - a bit more than 200 different aptamers, but they offer a selection of aptamers from scratch to meet your needs. Finally, one of most well-known companies offering aptamer-based products is SomaLogic. Their SOMAscan platform (https://somalogic.com/technology/our-platform/) based on SOMAmers is designed for multiplex detection of protein biomarkers in blood serum as a biochip. By now more than 300,000 samples have been examined in research tests with the use of this platform for determining of more than 50 different diseases, including oncological, metabolic, infectious and degenerative diseases.

The market of aptamers is growing quite slowly. Still today they have their own niche and compete with antibodies in various applications. The basic methods of detection with the use of antibodies and with the use of aptamers will be discussed further.

Flow Cytofluorimetry and Fluorescence Imaging

Flow cytometry is a bioanalytical method used in both research and clinical diagnostics to detect the expression of markers on the cell surface and the intracellular components, to characterize and isolate cell populations, to analyze the number of cells, their size and granularity, to study cell proliferation and the cell cycle, to analyse the transgenic populations and to diagnose diseases. Flow cytometry methods allow us to characterize complex multicomponent cell compounds simultenously across several parameters, with each single cell in the mixture being analyzed separately. The speed of such process is up to several tens of thousands of events per second. Imaging of tissues, cells and organelles by means of fluorescence microscopy is also widely used; it allows, among other things, to evaluate the processes occurring in cells *in vivo*, and carry out the histochemical analysis. Low molecular weight fluorescent dyes (7-AAD, PI, etc.), fluorescent proteins (GFP, YFP, mCherry, etc.) can be used for imaging. But the main method of specific staining is antibodies labeled with various fluorescent markers. These antibodies are commercially available in a big variety. Despite the dominant use of antibodies in cytometry and fluorescence imaging nowadays, aptamers show their advantages in many studies. When modified by fluorescent dyes, antibodies can change their properties and lose their affinity. Therefore, fluorescent imaging of a target often requires the use of a sandwich of a specific antibody and a fluorophore-labeled anti-antibody still increasing the cost of an expensive analysis. It was estimated that the cost of one cytometric test with the use of aptamers is many times less costly than a similar test with the use of antibodies [9].

The first use of aptamers in cytometry was reported in 1996. The representatives of Becton Dickinson Immunocytometry Systems and NeXstar Pharmaceuticals Inc. published the results of the use of an aptamer against human neutrophil elastase (HNE) with its comparison with the use of monoclonal antibodies against the same target. The authors concluded that the obtained data are comparable and that there are advantages of using aptamers in cytometry, such as their availability and simplicity of introduction of a fluorescent label into the aptamer, in particular, and the possibility to optimize the structure of the resulting conjugate. They aslo stated the preferred use of oligonucleotide ligands for visualising the intracellular targets as they are much smaller than antibodies [10].

The obtained results were quite promising. Since then, there have been regular reports of successful use of aptamers in flow cytometry and FACS, and in fluorescence microscopic imaging. The anti-CD4 marker aptamer was one of first aptamers used for cytometric phenotyping of leukocytes. The obtained aptamer specifically stained cells with human CD4 and did not bind the control cells expressing mouse CD4 whose extracellular domain is 55% identical to the human one [11]. The proteins of cluster of differentiation are now most popular targets for obtaining aptamers already selected for a variety of CD markers. This data is quite fully summarized in Nozari and Berezovski review [12]. Such interest is obvious as anti-CD specific molecules, mainly antibodies, are used in cell biology and medicine, in diagnostics and therapy [13].

Aptamers for markers of various tumors including lymphomas, adenocarcinomas, small-cell lung cancer, prostate cancer, lung cancer, brain cancer and others have been selected and tested in a variety of experiments [14, 15]. Aptamers are widely used for the diagnostics of malignant diseases with the use of such methods as flow cytometry, histochemical analysis of biopsy specimens, and vital staining of tumors to obtain fluorescent images, including CT and MRI [16]. The small size of aptamers contribute to their more effective penetration into all cancer cells, including the tumor core, while the unbound molecules are rapidly excreted by the kidneys without lingering in the body. This is important for *in vivo* diagnostics [17, 18]. Apart from that, aptamers show a high signal-to-background ratio that exceeds such ratio for antibodies. This ability can make aptamers leaders in tumor diagnostics [18, 19].

Aptamers are also used for cytometric and fluorescence detection of pathogenic microorganisms and viruses in food products, in water and clinical samples [14, 20, 21]. When the flow cytometry method is used for detection of pathogens, it is convenient to use nanoparticles conjugated with specific aptamers, for example, silica nanoparticles [22], gold nanoparticles [23], quantum dots [24] or other carriers could also be used for this purpose. Researchers have reported the high specificity and sensitivity of such process of detection. Thus, it was possible to obtain an aptamer able to detect *Bifidobacterium breve* in a mixture of closely related microorganisms with the sensitivity of 1,000 cfu/ml [25]. The sensitivity data of the bacteria detection with the use of a cytometer published by research groups varies within three orders of magnitude (from several dozens to 10^3 cfu/ml) and apparently strongly depends on the complexity of the sample under study. Detection of pathogenic microorganisms by cytometry with the use of aptamers is a simple and affordable method, but it is not always most effective. The use of the same aptamers in other methods sometimes allows to obtain better results. It is important to choose a more appropriate research method for each pathogen and each type of a sample [21].

BIOSENSORS

Biosensors are defined as independent analytical devices based on biological materials such as tissues, microorganisms, organelles, cell receptors, enzymes, antibodies, nucleic acids and other products of biological origin, as well as structures derived by biotechnology such as recombinant antibodies, recombinant proteins, aptamers, etc., or biomimics such as synthetic catalysts or combinatorial ligands [26]. Any biosensor consists of a biological recognition element (bioreceptor) and a transducer that recognizes and interprets the detected signal. The transducer detects the signal from the binding with the target molecule or, vice versa, loss of the signal due to the binding. The transducer can read electrochemical, optical, fluorescent, chemiluminescent, calorimetric, resonant (SPR) signals, etc. [27]. It is possible to combine all types of bioselecting elements with nearly any transducer. This gives a huge variety in types of biosensors. It must be noted that reliability, sensitivity and selectivity of a biosensor depend primarily on the bioreceptor that binds the analyte.

The use of antibodies in biosensors has been reported since mid-1970s [28, 29]. Antibody-based biosensors have been developed for detection of small contaminants in environmental objects [30, 31], of toxins [32, 33], of bacteria [33, 34] in mixed samples, etc. Polyclonal, monoclonal and recombinant antibodies [35] are used in these tests.

Biosensors based on aptamers (aptasensors) are not so widely used as other types of biosensors, but the work on developing such biosensors is quite active. Aptamers when applied in this procedure have some advantages over the use of antibodies. They are characterized by nearly similar affinity and specificity, but have greater physical and chemical stability. The development of biosensors for detection of low-molecular weight compounds and their metabolites is much easily carried out in the form of an aptasensor, since it is much simpler to obtain aptamers for such targets than for antibodies. Antibodies change their affinity under conditions that do not correspond to physiological ones, while aptamers can be selected in the conditions required for a certain assay. The simplicity of their modification contributes to constructing such an aptamer ligand that can be immobilized on any substrate in a strictly oriented manner. The ability to restore the structure after multiple cycles of denaturation/renaturation increases the lifetime of the aptasensor, and protection from nucleases can be provided just by competent post-modification.

One of the first to obtain was an aptasensor based on a biotinylated RNA aptamer against L-adenosine with a fiber-optic detection of a fluorescent signal from a FITC-labeled L-adenosine. High selectivity of L-adenosine binding with the sensitivity in the sub-micromolar range was achieved [36]. At about the same time, a highly effective, sensitive and rapid aptasensor was developed for the detection of thrombin. This aptasensor made it possible to directly detect at least 0.7 amol of thrombin in 140 pl of the sample in one step and 10 minutes. One slide covered with the used aptamer could be applied many times without losing its properties. The authors have discovered one more advantage of using aptamers - one can fluorescently label the ligand and record the signal change due to the analyte binding with it (in this work, fluorescent anisotropy was used to detect the signal). The use of a native unlabeled analyte significantly reduced the sample preparation time [37].

By now the data on a variety of developed aptasensors have been published. There are publications about aptasensors for detection of low-molecular weight substances, for example, such as tetracycline (limit of detection of 0.035 μg/l) [38], ATP (LoD 0.035 μM) [39], about various test systems for detection of thrombin, some of which have a limit of detection in the femtomolar range [40, 41], the system which detects cocaine in concentrations up to 1.34 pM in serum [42], etc. There are aptasensors for detection of pathogens and their markers such as HIV-1 TAT protein with a detection limit of 0.25 ppm (which is almost similar to the data [43] obtained using an immunosensor), H5N1 avian influenza virus in the samples taken from birds in concentrations from 0.128 to 12.8 HAU [44], *Bacillus thuringiensis* spores (LoD 10^3 cfu/ml) [45], *Escherichia coli* O157:H7 and *Salmonella typhimurium* bacteria (LoD 10^5 cfu\ml which is lower than or comparable to previously described immunosensors [46, 47]) [48].

Aptasensors are expected to play an important role in diagnostics, pharmacogenomics, medicine, forensic analysis, and biosafety [49]. Many of existing aptamers could be adapted for their use as bioreceptors. This can be specifically referred to such a rapidly developing group of substances as aptazymes [50, 51].

Replacement of Antibodies with Aptamers in Classical Immunoassay Systems

For long, immunological methods have been the "gold standard" of biomedical research. Monoclonal antibodies have high affinity and selectivity of interaction. But the above-mentioned disadvantages of antibodies, especially loss of specificity due to modification or denaturation, present a challenge of developing similar methods that use other types of ligands - long-lived, not so demanding for the conditions of the assay, easily adaptable to any task. The researchers have started to focus their attention at the use of synthetic ligands, including aptamers. Aptamers can be used as a solution or could be immobilized for the interaction on the surface of such solid carriers as gold or carbon nanoparticles, magnetic or glass beads, plastic and glass plates and slides, various resins and activated sorbents, graphene oxide, etc. [52]. Aptamers are applicable in all systems where they can completely replace antibodies or be combined with them.

Lateral Flow Immunoassay (LFIA) and Lateral Flow Aptamer Assay (LFAA)

Lateral flow assay (LFA) is one of the most convenient type of test systems - simple and quick in use, cheap in production, compact and portable, requiring no additional equipment or a highly skilled specialist. This test system exist in many variations as different bio-recognition molecules, labels and methods of detection can be used in its design. The test can be made in the classical "sandwich" format, in the format of competitive analysis or a multiplex test system. Colloidal gold, fluorescent or chemiluminescent labels, magnetic particles or enzymes are used as labels. Most of the existing LFA tests are based on immunoglobulins. When developing such a test system, the detection limit of the target analyte that determines the properties of a bio-recognizing substance is crucially important. For antibody-based test strips the detection limit has been reached within the range from nanomolar to picomolar [53].

Still a challenge remains to design lateral flow immunoassay (LFIA) tests for the detection of small molecules. This is mainly due to the difficulty of obtaining antibodies against small and non-immunogenic targets and the inaccessibility of binding sites. Aptamers become indispensable in such cases as it is possible to select them for almost any target. LFA in the format of an immunochromatographic test is often carried out in a sandwich binding assay. This can be easily applied to detect large molecules, but is not suitable for detecting small analytes. In this case, there is only one binding site and it is completely overlapped in binding with the first antibody not leaving a chance for the formation of a fully functional immuno-sandwich. This can be solved by the use of split aptamers or a competitive analysis format. In the lateral flow aptamer assay (LFAA) aptamers were successfully applied to the C-reactive protein and ampicillin in a competitive analysis format in order to detect ampicillin in milk. The authors have discovered the cross-reactivity of the aptamer to ampicillin with the C-reactive protein, which served as basis for creation of the test that determine up to 185 mg/l of the antibiotic. However, the discovered cross-reactivity suggests that interference with other substances can also occur, therefore any obtained aptamer should be tested on a fairly wide panel of possible targets [54]. Fluorescence detection was used to detect ochratoxin A with a sensitivity of 1.9 ng/ml, which is comparable to the sensitivity of an enzyme-linked immunosorbent assay (ELISA). But conducting the test in the LFAA format takes 15 minutes only and does not use the devices required for carrying out ELISA [55]. Up to 20 μM of ATP in a urine sample was detected using a competitive method where ATP and a probe complementary to the aptamer segment competed with a specific aptamer [56]. The use of aptamers in LFAA for detection of aflatoxin B1 made it possible to reach a detection limit of 0.1 ng/ml. This LoD is lower than in LFIA test (0.5 ng/ml), but at the same time is significantly higher than LoD in the classical ELISA (4 pg/ml). The authors note that despite the relatively low detection limit, there are obvious advantages of LFIA and LFAA tests: saved time (15-30 minutes versus 2 hours for the ELISA) and ease of performance (a minimum number of steps is required for the assay) [57].

Was shown a possibility of successful multiplex detection by LFAA test simultaneously across three targets completely different in size, structure, nature in the format of a fluorescent competitive test for the detection of mercury ions (LoD 5 ppb), ochratoxin A (LoD 3 ng/ml) and *Salmonella* (LoD 85 cfu/ml) without cross-reactions with other ions, small molecules and bacteria [58].

A variety of LFAA tests have also been developed for the detection of bacteria, viruses, their antigens and toxins. The use of quantum dots as a signal detection method in the LFAA test made it possible to increase the detection limit of *E. coli* O157:H7, *Listeria monocytogenes* and *Salmonella enterica* bacteria (300-600 cfu versus 3000-6000 cfu per test) by an order of magnitude compared with the tests where the colloidal gold based test was used [59]. The combination of an aptamer and an antibody allowed to detect the cholera toxin in a concentration up to 0.6 ng/ml (or 1 ng/ml when was visually evaluated the test results) [60]. Obtaining a pair of ligands specific for H5N2 avian influenza virus and using them in a two-site LFAA allowed to detect up to 2.09×10^5 EID_{50}/ml in guano samples. This is not worse than or even exceeds the kits for detection of avian influenza virus commercially available today [61]. One of the main advantages, namely the possibility of aptamer amplification, was used in the test for the detection of *Salmonella enteritidis* (LoD 10^1 cfu). Two aptamers specific to surface proteins of the *S. enteritidis* were used, one allowed the sample to be concentrated by binding with the surface of the magnetic bead, and the other served as a target for isothermal amplification. The resulting amplicons were analyzed in LFAA [62].

Thus, aptamers are not only successfully used in the LFA test systems, but also allow to achieve better results. LFA tests using target capture by antibodies require certain test conditions. Most tests require special temperature conditions during transportation (cold chain is required) and test conditions (tests must be heated to the optimal temperature of use). The higher stability of aptamers allows to simplify storage and transportation conditions. The advantage of using aptamers for the analysis of samples that differ in composition and conditions from physiological ones is also obvious. Most LFA tests use the two-site or "sandwich" binding so the resulting structure may be too massive for detection of small analytes, while the aptamers are significantly smaller. One can use a pair of two different aptamers or split-aptamers separated into two binding sequences [63, 64], which means that it is possible to design a "sandwich" system even with one aptamer sequence. The simplicity and accessibility of the modification of the nucleotide sequence and the small size of aptamers allows to make the post-modification changes that give a clearly oriented immobilization of the ligand and a tight fit of functionally active aptamers. The antibodies after modifications can change their affinity, they are more difficult to attach to the surface directionally, they also take up more space, and, as a result, most part of the space could be covered with antibodies inaccessible for interaction making the surface nonfunctional. This could become a key aspect in the design of test strips. We could also discuss the development of a multi-use LFA test with the regeneration of an aptamer-coated surface [65].

ELISA-Like Methods Using Ligand Amplification: AptaPCR and iaPCR

Probably the most obvious advantage of aptamers over antibodies is the possibility of their amplification both by classical PCR methods and, for example, using an isothermal amplification reaction [62, 66]. This manyfold enhances the signal from each individual aptamer entered into reaction with the target. The traditional use of polymerase chain reaction involves amplifying the target DNA directly, and if it is necessary to detect targets of a non-nucleotide nature without losing the sensitivity of the test, two approaches can be used. We can either improve the instrumentation base for detecting ultra-small signals or many times enhance the signal from a reporter associated with the target with the use of widely spread laboratory instruments.

Back in the 90s of the 20th century, the second of the above-mentioned approaches was applied in immuno-PCR [67]. Immuno-PCR is based on the classical sandwich assay using two antibodies on the ELISA principle. In this case, the binding signal is detected not by an enzymatic color reaction or a fluorescent or chemiluminescent signal, but by a DNA reporter associated with the detecting antibody. DNA reporter allows to visualize the binding signal in PCR with the product detection in agarose gel or in RT-PCR. In both cases, each molecule of the reacted reporter is copied many times which contributes to fixing the signal even from the minimum amount of the target captured, and therefore increases the sensitivity of the test.

However, the development of immuno-PCR test systems has some difficulties that have to be overcome in an each performed elaboration. The main problem is the process of conjugation of the antibody and the reporter which is almost impossible to be made receiving the same ratio of antibody/DNA for each molecule and in each iteration of conjugation as a result. That makes the quantification of the analyte difficult as, in such case, each single antibody molecule conjugated to the reporter gives its own signal level poorly comparable with the signal from other molecules [68]. In addition, the non-specific binding of the DNA reporter with the components of the reaction and off-targets presents a problem. As a result, we have a sufficiently high background signal and false-positive results may occur.

Immuno-PCR and its modifications were used to detect viruses, bacteria, protozoa and their antigens, prion proteins, bacterial toxins, mycotoxins, etc. [69]. However, this technology still lacks sufficient automatization, it requires reduction in the number of steps and assay duration, the number of components in the test and, finally, reduction of the cost.

Aptamer-PCR (aptaPCR) or immuno-aptamer PCR (iaPCR) with a proper approach to the development of the test system allows to reduce the background signal as the ligands, including nucleotide ones, bind to the target site accordingly. To reduce or eliminate the background signal, it is possible to carry out the required number of counter-selection steps with obtaining an aptamer with low cross-reactivity under the specific test conditions. Besides, using the aptaPCR or iaPCR approach can reduce the number of components involved in the test, because when the aptamer is used, the detecting ligand by itself is a signal amplified molecule and does not require additional labeling of the obtained detecting complex. When the separation of the aptamer from the detecting sandwich is required, this can be easily and effectively performed with the use of simple temperature denaturation.

AptaPCR based on a two-aptamer sandwich with signal detection in RT-PCR format was first reported in a paper about the detection of thrombin where the limit of detection of 450 fM was reached. It was 20,000 times higher than the result for a sandwich analysis based on the aptamer capture with enzymatic detection [70]. To date, there have been advances in the development of an aptaPCR test system for the detection of a wide range of targets. Thus, apta-qPCR with high specificity have been developed, it can detect up to 146.7 fM of staphylococcal enterotoxin type A (SEA) distinguishing it from type D and type E toxins (with the homology degree of 50% and 83% respectively) [71]. AptaPCR is used to detect pathogens in food, for example, *Salmonella typhimurium* bacteria in the sample of turkey meat. The test was carried out with the refinement of the sample by magnetic separation. The detection limit in the clean sample was 10^2 cfu/ml, in the test of turkey meat - 10^3 cfu/ml [72]. This is not the best result for an aptamer detection system as the impedance-aptasensor capable to detect up to 1×10^2 cfu/ml in apple juice has already been obtained [73]. The aptaPCR test system for the detection of food allergens, such as β-conglutin (LoD 85 pM = 25 ng/ml), has also been described. The allergen was detected in lupine pollen samples using the competitive aptaPCR without observable cross-reactivity with some other possible targets [74]. The developed test system for the detection of the H9N2 influenza virus shows the results with a limit of detection of 1.00E + 2 $TCID_{50}$/ml (including clinical samples), which is 1,000 times less than the limit of detection of a commercially available ELISA kit [75]. AptaPCR is also applicable for the detection of cancer cells. Thus, the split-aptaPCR system was used to diagnose liver cancer, up to 100 target cells were detected in the test [76].

The analysis under the general name of aptaPCR exists in various modifications. These are aptaPCR with circular amplification [77], with caged probes [78, 79], with the use of temperature-dependent split aptamers [76], etc. This area is developing rapidly, and more formats and interpretations of aptaPCR systems are sure to appear soon.

Aptamers are used in ELISA-like test systems not only as a specific ligand that directly detects a target. It has also been suggested that an aptamer can be applied to the Fc fragment of rabbit immunoglobulin G as a reporter in iaPCR, while the binding sandwich was represented by pair of antibodies and the reporter was represented by an aptamer. Such use of the aptamer made it possible to increase the efficacy of analyte detection from 10 pg/ml in ELISA to 100 fg/ml in iaPCR. The use of a specific DNA reporter significantly reduces the risks of false-positive results due to non-specific binding of the DNA label [80].

Aptamers are also used in ELISA-like techniques such as Enzyme-Linked OligoNucleotide Assay (ELONA) [81] and Enzyme-Linked Aptamer Sorbent Assay (ELASA) [82] that in their original variant do not use signal amplification by the PCR. These methods use all the advantages of the affinity, stability and reproducibility of aptamers, and solve the main problem of highly sensitive techniques, such as likely high levels of background signal in case of a non-optimal aptamer used. Certainly, such techniques can be equal in sensitivity only to ELISA, sometimes slightly exceeding it.

Conclusion

Market research on aptamers suggests an expected increase in its volume by 20% annually. According to the forecast by Transparency Market Research, the size of aptamers market will reach $5 billion by the year 2025 [83]. It means there is a growing interest in this topic.

So, why have aptamers not yet replaced antibodies, have not become a routine reagent, and the commercial offer is so small, though growing every year?

Researchers have developed dozens of ways of aptamer selection, but there is neither universal protocol or standardized selection methods. As discussed above, commercially available kits for carrying out the selection have just started to emerge, the offer is extremely small, and as there is no sufficiently wide experience of their use, the universal use of such kits is still being discussed. In theory, highly specific aptamers can be obtained for any target. But when we have a complex target, then there are more targets homologous to it and more possible non-homologous analytes in the studied samples of a complex composition. So the more strict requirements are placed on the ligand, then the selection with counter-selection steps included is more difficult, and the test of the original and modified aptamer is more complex. Naturally, antibodies can also have cross-reactivity and, unlike aptamers, their specificity cannot be influenced. On the other hand, the more sensitive the test system should be (especially for methods involving the amplification stage of aptamers), the more strictly the ligand selection is made. Any minor cross reaction will significantly reduce the sensitivity of the resulting test system.

Uncertainty in the choice of selection methods and making modifications to the nucleotide sequence of the aptamer also leads to the fact that it is easier for researchers to use or improve already existing, well-described and tested sequences than to develop their own approaches to obtaining a good-quality aptamer ligand. One of most commonly used aptamers is a well-studied antithrombin aptamer. Hundreds of researchers have used it in their research work, they preferred to focus on different ways of using it for detecting a single target, rather than get aptamers and test systems based on them for other clinically significant markers. This effect is known as the "thrombin problem" [12, 84].

The problem of the adaptation of aptamers for use in the test of complex multicomponent mixtures, or real clinical specimens, of the samples of environmental objects or food products still remains. It has been shown that the conformation and, therefore, the binding efficacy of an aptamer with a target is affected by differences in selection conditions of a particular aptamer and conditions of a real test (buffer composition, temperature, pH, etc.) [85]. This means that before the selection, it is worth thinking about further application of the obtained sequence and keeping in mind the real conditions of the analyte detection in a certain test format.

It is also worth taking into account that the use of antibodies is currently just more traditional and familiar to most users, and companies produce and promote a huge amount of antibodies suitable for almost any method of use, even though with some restrictions.

Aptamers as an alternative tool of detection still has a lot of advantages. The use of aptamers made it possible to accurately detect the targets that were previously inaccessible for effective detection with use of antibodies. Among such targets we can mention toxic substances, many small molecules, metabolites and non-immunogenic molecules. With the help of aptamers, it became possible to distinguish certain cell types that were previously indistinguishable, especially with respect to the diagnostics of oncological diseases. Now we can identify cancer cell subtypes [14]. The stability of modified aptamers makes it possible to produce multiuse diagnostic systems with long storage periods that do not lose the efficacy of target recognition under various conditions.

Today, not all developers possess the tools and knowledge for obtaining aptamers, this technology is known to a relatively narrow circle of scientists and researchers. Further development of the aptamer technology and expansion of its zone of influence will certainly lead to a considerable decrease in the cost of any systems that use aptamers in their composition.

Acknowledgments

This work was supported by the Sectoral Scientific Program of the Russian Federal Service for Surveillance on Consumer Rights Protection and Human Wellbeing.

References

[1] Köhler, Georges, and Cesar Milstein. 1975."Continuous cultures of fused cells secreting antibody of predefined specificity." *Nature* 256.5517: 495. doi:10.1038_256495a0.

[2] Tuerk, Craig, and Larry Gold. 1990. "Systematic evolution of ligands by exponential enrichment: RNA ligands to bacteriophage T4 DNA polymerase." *Science* 249.4968: 505-510. doi:10.1126/science. 2200121.

[3] Ruscito, Annamaria, and Maria C. DeRosa. 2016. "Small-molecule binding aptamers: Selection strategies, characterization, and

applications." *Frontiers in Chemistry* 4: 14. doi:10.3389/fchem.2016.00014.

[4] Klussmann, Sven, ed. 2006. *The aptamer handbook: functional oligonucleotides and their applications*. John Wiley & Sons.

[5] White, Rebekah R., Bruce A. Sullenger, and Christopher P. Rusconi. 2000. "Developing aptamers into therapeutics." *The Journal of Clinical Investigation* 106.8: 929-934. doi: 10.1172/jci11325.

[6] Ni, Shuaijian, Houzong Yao, Lili Wang, Jun Lu, Feng Jiang, Aiping Lu and Ge Zhang. 2017. "Chemical modifications of nucleic acid aptamers for therapeutic purposes." *International journal of molecular sciences* 18.8: 1683. doi: 10.3390/ijms18081683.

[7] Hasegawa, Hijiri, Nasa Savory, Koichi Abe and Kazunori Ikebukuro. 2016. "Methods for improving aptamer binding affinity." *Molecules* 21.4: 421. doi:10.3390/molecules21040421.

[8] Virgilio, Antonella, Teresa Amato, Luigi Petraccone, Francesca Esposito, Nicole Grandi, Enzo Tramontano, Raquel Romero et al. 2018. "Improvement of the activity of the anti-HIV-1 integrase aptamer T30175 by introducing a modified thymidine into the loops." *Scientific Reports* 8.1: 7447. doi: 10.1038_s41598-018-25720-1.

[9] Zhang, Peng, Nianxi Zhao, Zihua Zeng, Chung-Che Chang and Youli Zu. 2010. "Combination of an aptamer probe to CD4 and antibodies for multicolored cell phenotyping." *American Journal of Clinical Pathology* 134.4: 586-593. doi:10.1309/AJCP55KQYWSGZRKC.

[10] Davis, Kenneth A., Barnaby Abrams, Yun Lin and Sumedha D. Jayasena. 1996. "Use of a high affinity DNA ligand in flow cytometry." *Nucleic Acids Research* 24.4: 702-706. doi:10.1093/nar/24.4.702.

[11] Davis, Kenneth A., Barnaby Abrams, Yun Lin and Sumedha D. Jayasena. 1998. "Staining of cell surface human CD4 with 2′-F-pyrimidine-containing RNA aptamers for flow cytometry." *Nucleic Acids Research* 26.17: 3915-3924. doi:10.1093/nar/26.17.3915.

[12] Nozari, Amin, and Maxim V. Berezovski. 2017. "Aptamers for CD antigens: from cell profiling to activity modulation." *Molecular Therapy-Nucleic Acids* 6: 29-44. doi:10.1016/j.omtn.2016.12.002.

[13] Heddy, Z. O. L. A., and Bernadette Swart. 2005. "The human leucocyte differentiation antigens (HLDA) workshops: the evolving role of antibodies in research, diagnosis and therapy." *Cell Research* 15.9: 691. doi:10.1038_sj.cr.7290338.

[14] Meyer, Michael, Thomas Scheper, and Johanna-Gabriela Walter. 2013. "Aptamers: versatile probes for flow cytometry." *Applied Microbiology and Biotechnology* 97.16: 7097-7109. doi:10.1007/s00253-013-5070-z.

[15] Sun, Hongguang, Weihong Tan, and Youli Zu. 2016. "Aptamers: versatile molecular recognition probes for cancer detection." *Analyst* 141.2: 403-415. doi: 10.1039/C5AN01995H.

[16] Wu, Xu, Jiao Chen, Min Wu, and Julia Xiaojun Zhao. 2015. "Aptamers: active targeting ligands for cancer diagnosis and therapy." *Theranostics* 5.4: 322. doi:10.7150/thno.10257.

[17] Wang, Andrew Z., and Omid C. Farokhzad. 2014. "Current progress of aptamer-based molecular imaging." *Journal of Nuclear Medicine* 55.3: 353-356. doi:10.2967/jnumed.113.126144.

[18] Xiang, Dongxi, Sarah Shigdar, Greg Qiao, Tao Wang, Abbas Z. Kouzani, Shu-Feng Zhou, Lingxue Kong et al. 2015. "Nucleic acid aptamer-guided cancer therapeutics and diagnostics: the next generation of cancer medicine." *Theranostics* 5.1: 23. doi:10.7150/thno.10202.

[19] Kaur, Harleen, Jasmine J. Li, Boon-Huat Bay, and Lin-Yue Lanry Yung. 2013. "Investigating the antiproliferative activity of high affinity DNA aptamer on cancer cells." *PloS one* 8.1: e50964. doi: 10.1371/journal.pone.0050964.

[20] Hamula, Camille L. A., Hongquan Zhang, Feng Li, Zhixin Wang, X. Chris Le, and Xing-Fang Li. 2011. "Selection and analytical applications of aptamers binding microbial pathogens." *TrAC Trends in Analytical Chemistry* 30.10: 1587-1597. doi:10.1016/j.trac.2011.08.006.

[21] Alizadeh, Naser, Mohammad Yousef Memar, Seyyed Reza Moaddab, and Hossein Samadi Kafil. 2017. "Aptamer-assisted novel

technologies for detecting bacterial pathogens." *Biomedicine & Pharmacotherapy* 93: 737-745. doi:10.1016/j.biopha.2017.07.011.

[22] He, Xiaoxiao, Yuhong Li, Dinggen He, Kemin Wang, Jingfang Shangguan, and Hui Shi. 2014. "Aptamer-fluorescent silica nanoparticles bioconjugates based dual-color flow cytometry for specific detection of Staphylococcus aureus." *Journal of Biomedical Nanotechnology* 10.7: 1359-1368. doi:10.1166/jbn.2014.1828.

[23] Yuan, Jinglei, Shijia Wu, Nuo Duan, Xiaoyuan Ma, Yu Xia, Jie Chen, Zhansheng Ding, and Zhouping Wang. 2014. "A sensitive gold nanoparticle-based colorimetric aptasensor for Staphylococcus aureus." *Talanta* 127: 163-168. doi:10.1016/j.talanta.2014.04.013.

[24] Duan, Nuo, Shijia Wu, Ye Yu, Xiaoyuan Ma, Yu Xia, Xiujuan Chen, Yukun Huang, and Zhouping Wang. 2013. "A dual-color flow cytometry protocol for the simultaneous detection of Vibrio parahaemolyticus and Salmonella typhimurium using aptamer conjugated quantum dots as labels." *Analytica chimica acta* 804: 151-158. doi:10.1016/j.aca.2013.09.047.

[25] Hu, Lujun, Linlin Wanga, Wenwei Luab, Qixiao Zhaiab, Daming Fanab, Xiaoming Liuab, Jianxin Zhao et al. 2017. "Selection, identification and application of DNA aptamers for the detection of Bifidobacterium breve." *RSC Advances* 7.19: 11672-11679. doi:10.1039/C6RA27672E.

[26] Lazcka, Olivier, F. Javier Del Campo, and F. Xavier Munoz. 2007. "Pathogen detection: a perspective of traditional methods and biosensors." *Biosensors and Bioelectronics* 22.7: 1205-1217. doi: 10.1016/j.bios.2006.06.036.

[27] Iqbal, Shahzi S., Michael W. Mayo, John G. Bruno, Burt V. Bronk, Carl A. Batt, and James P. Chambers. 2000. "A review of molecular recognition technologies for detection of biological threat agents." *Biosensors and Bioelectronics* 15.11-12: 549-578. doi:10.1016/S0956-5663(00)00108-1.

[28] Aizawa, Masuo, Aya Morioka, Hideaki Matsuoka, Shuichi Suzuki, Yoichi Nagamura, Rikio Shinohara, and Isao Ishiguro. 1976. "An

enzyme immunosensor for IgG." *Journal of Solid-Phase Biochemistry* 1.4: 319-328. doi:10.1007/BF02990970.

[29] Thomas, J. D. R., Masuo Aizawa, I. J. Higgins, and W. J. Albery. 1987. "Immunosensors." Philosophical Transactions of the Royal Society of London. Series B, *Biological Sciences*: 121-134.

[30] Tschmelak, Jens, Guenther Proll, and Guenter Gauglitz. 2004. "Sub-nanogram per litre detection of the emerging contaminant progesterone with a fully automated immunosensor based on evanescent field techniques." *Analytica chimica acta* 519.2: 143-146. doi:10.1016/j.aca.2004.06.031.

[31] Zhang, Zhen, Kun Zeng, and Jingfu Liu. 2017. "Immunochemical detection of emerging organic contaminants in environmental waters." *TrAC Trends in Analytical Chemistry* 87 : 49-57. doi:10.1016/j.trac.2016.12.002.

[32] Ligler, Frances S., Chris Rowe Taitt, Lisa C. Shriver-Lake, Kim E. Sapsford, Yura Shubin, and Joel P. Golden. 2003. "Array biosensor for detection of toxins." *Analytical and Bioanalytical Chemistry* 377.3: 469-477. doi: 10.1007/s00216-003-1992-0.

[33] Sharma, Harsh, and Raj Mutharasan. 2013. "Review of biosensors for foodborne pathogens and toxins." *Sensors and Actuators B: Chemical* 183: 535-549. doi:10.1016/j.snb.2013.03.137.

[34] Vaisocherová-Lísalová, Hana, Ivana Víšová, Maria Laura Ermini, Tomáš Špringer, Xue Chadtová Song, Jan Mrázek, Josefína Lamačová et al. 2016. "Low-fouling surface plasmon resonance biosensor for multi-step detection of foodborne bacterial pathogens in complex food samples." *Biosensors and Bioelectronics* 80: 84-90. doi:10.1016/j.bios.2016.01.040.

[35] Hock, Bertold., Martin Seifert, and Karl J. Kramer. 2002. "Engineering receptors and antibodies for biosensors." *Biosensors and Bioelectronics* 17.3: 239-249. doi:10.1016/S0956-5663(01)00267-6.

[36] Kleinjung, F., S. Klussmann, V. A. Erdmann, F. W. Scheller, J. P. Fürste, and F. F. Bier. 1998. "High-affinity RNA as a recognition element in a biosensor." *Analytical Chemistry* 70.2 (1998): 328-331. doi:10.1021/ac9706483.

[37] Potyrailo, Radislav A., Richard C. Conrad, Andrew D. Ellington, and Gary M. Hieftje. 1998. "Adapting selected nucleic acid ligands (aptamers) to biosensors." *Analytical Chemistry* 70.16: 3419-3425. doi:10.1021/ac9802325.

[38] Alawad, Ahmad, Georges Istamboulié, Carole Calas-Blanchard, and Thierry Noguer. 2019. "A reagentless aptasensor based on intrinsic aptamer redox activity for the detection of tetracycline in water." *Sensors and Actuators B: Chemical* 288: 141-146. doi:10.1016/j.snb.2019.02.103.

[39] Xu, Peipei, and Guangfu Liao. 2018. "A Novel Fluorescent Biosensor for Adenosine Triphosphate Detection Based on a Metal–Organic Framework Coating Polydopamine Layer." *Materials* 11.9: 1616. doi:10.3390/ma11091616.

[40] Xiang, Yun, Mingyi Xie, Ralph Bash, Julian J. L. Chen, and Joseph Wang. 2007. "Ultrasensitive Label-Free Aptamer-Based Electronic Detection." *Angewandte Chemie International Edition* 46.47: 9054-9056. doi:10.1002/anie.200703242.

[41] Deng, Chunyan, Jinhua Chen, Zhou Nie, Mengdong Wang, Xiaochen Chu, Xiaoli Chen, Xilin Xiao et al. 2008. "Impedimetric aptasensor with femtomolar sensitivity based on the enlargement of surface-charged gold nanoparticles." *Analytical Chemistry* 81.2: 739-745. doi:10.1021/ac800958a.

[42] Oueslati, Rania, Cheng Cheng, Jayne Wu, and Jiangang Chen. 2018. "Highly sensitive and specific on-site detection of serum cocaine by a low cost aptasensor." *Biosensors and Bioelectronics* 108: 103-108. doi:10.1016/j.bios.2018.02.055.

[43] Minunni, Maria, Sara Tombelli, Antonella Gullotto, Ettore Luzi, and Marco Mascini. 2004. "Development of biosensors with aptamers as bio-recognition element: the case of HIV-1 Tat protein." *Biosensors and Bioelectronics* 20.6: 1149-1156. doi:10.1016/j.bios.2004.03.037.

[44] Bai, Hua, Ronghui Wang, Billy Hargis, Huaguang Lu, and Yanbin Li. 2012. "A SPR aptasensor for detection of avian influenza virus H5N1." *Sensors* 12.9: 12506-12518. doi:10.3390/s120912506.

[45] Ikanovic, Milada, Walter E. Rudzinski, John G. Bruno, Amity Allman, Maria P. Carrillo, Sulatha Dwarakanath, Suneetha Bhahdigadiet al. 2007. "Fluorescence assay based on aptamer-quantum dot binding to Bacillus thuringiensis spores." *Journal of Fluorescence* 17.2: 193-199. doi:10.1007/s10895-007-0158-4.

[46] Yang, Liju, Yanbin Li, and Gisela F. Erf. 2004. "Interdigitated array microelectrode-based electrochemical impedance immunosensor for detection of Escherichia coli O157:H7." *Analytical Chemistry* 76.4: 1107-1113. doi:10.1021/ac0352575.

[47] Wang, Yixian, Zunzhong Ye, Chengyan Si, and Yibin Ying. 2011. "Subtractive inhibition assay for the detection of E. coli O157: H7 using surface plasmon resonance." *Sensors* 11.3: 2728-2739. doi:10.3390/s110302728.

[48] Wu, Wen-he, Min Li, Yue Wang, Hou-Xian Ouyang, Lin Wang, Ci-Xiu Li, Yu-Chen Cao et al. 2012. "Aptasensors for rapid detection of Escherichia coli O157: H7 and Salmonella typhimurium." *Nanoscale Research Letters* 7.1: 658. doi:10.1186/1556-276X-7-658.

[49] Lim, Y. C., A. Z. Kouzani, and W. Duan. 2010. "Aptasensors: a review." *Journal of Biomedical Nanotechnology* 6.2: 93-105. doi:10.1166/jbn.2010.1103.

[50] Liu, Juewen, and Yi Lu. 2004. "Adenosine-dependent assembly of aptazyme-functionalized gold nanoparticles and its application as a colorimetric biosensor." *Analytical Chemistry* 76.6: 1627-1632. doi:10.1021/ac0351769.

[51] O'Sullivan, Ciara K. 2002. "Aptasensors – the future of biosensing?" *Analytical and Bioanalytical Chemistry* 372.1: 44-48. doi:10.1007/s00216-001-1189-3.

[52] Nezlin, Roald. 2016. "Use of aptamers in immunoassays." *Molecular Immunology* 70: 149-154. doi:10.1016/j.molimm.2015.12.009.

[53] Sajid, Muhammad, Abdel-Nasser Kawde, and Muhammad Daud. 2015. "Designs, formats and applications of lateral flow assay: A literature review." *Journal of Saudi Chemical Society* 19.6: 689-705. doi:10.1016/j.jscs.2014.09.001.

[54] Kaiser, Lars, Julia Weisser, Matthias Kohl, and Hans-Peter Deigner. 2018. "Small molecule detection with aptamer based lateral flow assays: Applying aptamer-C-reactive protein cross-recognition for ampicillin detection." *Scientific Reports* 8.1: 5628. doi:10.1038_ s41598-018-23963-6.

[55] Wang, Libing, Wei Chen, Wenwei Ma, Liqinag Liu, Wei Ma, Yuan Zhao, Yingyue Zhu et al. 2011. "Fluorescent strip sensor for rapid determination of toxins." *Chemical Communications* 47.5: 1574-1576. doi:10.1039/C0CC04032K.

[56] Zhang, Qiang, Hongbin Qiu, Fangqiang Tang, Ye Tao, Baosheng Guan, Xuechen Li, Wei Yang. 2016. "Aptamer-based dry-reagent strip biosensor for detection of small molecule ATP." *Chemistry Letters* 45.3: 289-290. doi:10.1246/cl.151077.

[57] Shim, Won-Bo, Min Jin Kim, Hyoyoung Mun, and Min-Gon Kim. 2014. "An aptamer-based dipstick assay for the rapid and simple detection of aflatoxin B1." *Biosensors and Bioelectronics* 62: 288-294. doi:10.1016/j.bios.2014.06.059.

[58] Jin, Birui, Yexin Yang, Rongyan He, Yong Il Park, Aeju Lee, Dan Bai, Fei Li et al. 2018. "Lateral flow aptamer assay integrated smartphone-based portable device for simultaneous detection of multiple targets using upconversion nanoparticles." *Sensors and Actuators B: Chemical* 276: 48-56. doi:10.1016/j.snb.2018.08.074.

[59] Bruno, John. 2014. "Application of DNA aptamers and quantum dots to lateral flow test strips for detection of foodborne pathogens with improved sensitivity versus colloidal gold." *Pathogens* 3.2: 341-355. doi:10.3390/pathogens3020341.

[60] Frohnmeyer, Esther, Nadine Tuschel, Tobias Sitz, Cornelia Hermann, Gregor T. Dahl, Florian Schulz, Antje J. Baeumner and Markus Fischer. 2019. "Aptamer lateral flow assays for rapid and sensitive detection of cholera toxin." *Analyst* 144.5: 1840-1849. doi:10.1039/ C8AN01616J.

[61] Kim, Sang Hoon, Junho Lee, Bang Hyun Lee, Chang-Seon Song, and Man Bock Gu. 2019. "Specific detection of avian influenza H5N2 whole virus particles on lateral flow strips using a pair of sandwich-

type aptamers." *Biosensors and Bioelectronics*. doi:10.1016/j.bios. 2019.03.061.

[62] Fang, Zhiyuan, WeiWu, Xuewen Lu, and Lingwen Zeng. 2014. "Lateral flow biosensor for DNA extraction-free detection of salmonella based on aptamer mediated strand displacement amplification." *Biosensors and Bioelectronics* 56: 192-197. doi:10. 1016/j.bios.2014.01.015.

[63] Bai, Yunfeng, Feng Feng, Lu Zhao, Zezhong Chen, Haiyan Wangb and Yali Duana. 2014. "A turn-on fluorescent aptasensor for adenosine detection based on split aptamers and graphene oxide." *Analyst* 139.8: 1843-1846. doi:10.1039/C4AN00084F.

[64] Chen, Junhua, and Lingwen Zeng. 2013. "Enzyme-amplified electronic logic gates based on split/intactaptamers." *Biosensors and Bioelectronics* 42: 93-99. doi:10.1016/j.bios.2012.10.030.

[65] Chen, Ailiang, and Shuming Yang. 2015. "Replacing antibodies with aptamers in lateral flow immunoassay." *Biosensors and Bioelectronics* 71: 230-242. doi:10.1016/j.bios.2015.04.041.

[66] Li, Feng, Hongquan Zhang, Zhixin Wang, Ashley M. Newbigging, Michael S. Reid, Xing-Fang Li, and X. Chris Le. 2014. "Aptamers facilitating amplified detection of biomolecules." *Analytical Chemistry* 87.1: 274-292. doi:10.1021/ac5037236.

[67] Sano, Takeshi, Cassandra L. Smith, and Charles R. Cantor. 1992. "Immuno-PCR: very sensitive antigen detection by means of specific antibody-DNA conjugates." *Science* 258.5079 : 120-122. doi:10.1126/ science.1439758.

[68] Niemeyer, Christof M., Michael Adler, Bruno Pignataro, Steven Lenhert, Song, Chi Lifeng, Harald Fuchs, and Blohm Dietmar. 1999. "Self-assembly of DNA-streptavidin nanostructures and their use as reagents in immuno-PCR." *Nucleic Acids Research* 27.23: 4553-4561. doi:10.1093/nar/27.23.4553.

[69] Mehta, Promod K., Ankush Raj, Netra Pal Singh, and Gopal K. Khuller. "Detection of potential microbial antigens by immuno-PCR (PCR-amplified immunoassay)." *Journal of Medical Microbiology* 63.5: 627-641. doi:10.1099/jmm.0.070318-0.

[70] Pinto, Alessandro, M. Carmen Bermudo Redondo, V. Cengiz Ozalp, and Ciara K. O'Sullivan. 2009. "Real-time apta-PCR for 20000-fold improvement in detection limit." *Molecular BioSystems* 5.5: 548-553. doi:10.1039/B814398F.

[71] Sedighian, Hamid, Raheleh Halabian, Jafar Amani, Mohammad Heiat, Mohsen Amin, and Abbas Ali Imani Fooladi. 2018. "Staggered Target SELEX, a novel approach to isolate non-cross-reactive aptamer for detection of SEA by apta-qPCR." *Journal of Biotechnology* 286: 45-55. doi:10.1016/j.jbiotec.2018.09.006.

[72] Wang, Lijun, Ronghui Wang, Hong Wang, Michael Slavik, HuaWei, and Yanbin Li. 2017. "An aptamer-based PCR method coupled with magnetic immunoseparation for sensitive detection of Salmonella Typhimurium in ground turkey." *Analytical Biochemistry* 533: 34-40. doi:10.1016/j.ab.2017.06.010.

[73] Bagheryan, Zahra, Jahan-Bakhsh Raoof, Mohsen Golabi, Anthony P. F. Turner, and Valerio Beni. 2016. "Diazonium-based impedimetric aptasensor for the rapid label-free detection of Salmonella typhimurium in food sample." *Biosensors and Bioelectronics* 80: 566-573. doi:10.1016/j.bios.2016.02.024.

[74] Svobodova, Marketa, Teresa Mairal, Pedro Nadal, M. Carmen Bermudo, and Ciara K. O'Sullivan. 2014. "Ultrasensitive aptamer based detection of β-conglutin food allergen." *Food chemistry* 165: 419-423. doi:10.1016/j.foodchem.2014.05.128.

[75] Hmila, Issam, Manoosak Wongphatcharachai, Nacira Laamiri, Rim Aouini, Boutheina Marnissi, Marwa Arbi, Srinand Sreevatsan, and Abdeljelil Ghram. 2017. "A novel method for detection of H9N2 influenza viruses by an aptamer-real time-PCR." *Journal of Virological Methods* 243: 83-91. doi:10.1016/j.jviromet.2017.01.024.

[76] Tang, Jinlu, Xiaoxiao He, Yanli Lei, Hui Shi, Qiuping Guo, Jianbo Liu, Dinggeng He et al. 2017. "Temperature-responsive split aptamers coupled with polymerase chain reaction for label-free and sensitive detection of cancer cells." *Chemical Communications* 53.87: 11889-11892. doi:10.1039/C7CC06218D.

[77] Guo, Limin, Lihua Hao, and Qiang Zhao. 2016. "An aptamer assay using rolling circle amplification coupled with thrombin catalysis for protein detection." *Analytical and Bioanalytical Chemistry* 408.17: 4715-4722. doi:10.1007/s00216-016-9558-0.

[78] Pinto, Alessandro, Sabine Lennarz, Alexandre Rodrigues-Correia, Alexander Heckel, Ciara K. O'Sullivan, and Günter Mayer. 2011. "Functional detection of proteins by caged aptamers." *ACS Chemical Biology* 7.2: 360-366. doi:10.1021/cb2003835.

[79] Civit, Laia, Alessandro Pinto, Alexandre Rodrigues-Correia, Alexander Heckel, Ciara K. O'Sullivan, and Günter Mayera. 2016. "Sensitive detection of cancer cells using light-mediated apta-PCR." *Methods* 97: 104-109. doi:10.1016/j.ymeth.2015.11.018.

[80] Yoshida, Yoshihito, Katsunori Horii, Nobuya Sakai, Hiromi Masuda, Makio Furuichi, and Iwao Waga. 2009. "Antibody-specific aptamer-based PCR analysis for sensitive protein detection." Analytical and *Bioanalytical Chemistry* 395.4: 1089-1096. doi:10.1007/s00216-009-3041-0.

[81] Stoltenburg, Regina, Petra Krafčiková, Viktor Víglaský, and Beate Strehlitz. 2016. "G-quadruplex aptamer targeting Protein A and its capability to detect Staphylococcus aureus demonstrated by ELONA." *Scientific Reports* 6 : 33812. doi:10.1038/srep33812

[82] Park, Ji Hoon, Min Hyeok Jee, Oh Sung Kwon, Sun Ju Keum, and Sung Key Jang. 2013."Infectivity of hepatitis C virus correlates with the amount of envelope protein E2: development of a new aptamer-based assay system suitable for measuring the infectious titer of HCV." *Virology* 439.1: 13-22. doi:10.1016/j.virol.2013.01.014.

[83] Aptamer Market (Material - Nucleic acid Aptamer, Peptide Aptamer; Selection Technique - SELEX Technique and Others; Application - Research, Diagnostics, Therapeutics) - Global Industry Analysis, *Size, Share, Growth, Trends, and Forecast* 2017 – 2025.

[84] Baird, Geoffrey S. 2010. "Where are all the aptamers?" *American Journal of Clinical Pathology*. 134.4: 529-531. doi:10.1309/AJCPFU4CG2WGJJKS.

[85] Li, Na, Jessica N. Ebright, Gwendolyn M. Stovall, Xi Chen, Hong Hanh Nguyen, Amrita Singh, Angel Syrett and Andrew D. Ellington. 20019. "Technical and biological issues relevant to cell typing with aptamers." *Journal of Proteome Research* 8.5: 2438-2448. doi:10.1021/pr801048z.

INDEX

#

β

A

B

C

D

E

F

G

H

I

J

L

M

N

O

P

Q

R

S

T

V

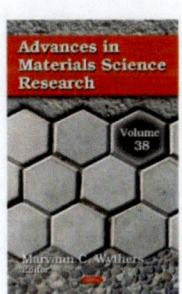

Related Nova Publications

MECHANISMS OF EVOLUTION

AUTHOR: Börje Ekstig

SERIES: Insights into Biological and Cultural Evolution

BOOK DESCRIPTION: In this book, we embark on an innovative, exploratory and interdisciplinary adventure, step by step following the author towards his quest of investigating evolution, its mechanisms, its direction and progress and, not least, the place of ourselves in it.

HARDCOVER ISBN: 978-1-53615-795-6
RETAIL PRICE: $160

To see a complete list of Nova publications, please visit our website at www.novapublishers.com